INTERNATIONAL RECENT ISSUES ABOUT ECDIS, e-NAVIGATION AND SAFETY AT SEA

International Recent Issues about ECDIS, e-Navigation and Safety at Sea

Marine Navigation and Safety of Sea Transportation

Editor

Adam Weintrit
Gdynia Maritime University, Gdynia, Poland

CRC Press
Taylor & Francis Group
Boca Raton London New York

CRC Press is an imprint of the
Taylor & Francis Group, an **informa** business

A BALKEMA BOOK

CRC Press/Balkema
P.O. Box 447, 2300 AK Leiden, The Netherlands
e-mail: Pub.NL@taylorandfrancis.com
www.crcpress.com – www.taylorandfrancis.com

First issued in hardback 2017

© 2011 Taylor & Francis Group, London, UK
CRC Press/Balkema is an imprint of Taylor & Francis Group, an Informa business

No claim to original U.S. Government works

ISBN-13: 978-0-4156-9112-3 (pbk)
ISBN-13: 978-1-1384-3581-0 (hbk)

Visit the Taylor & Francis Web site at
http://www.taylorandfrancis.com

and the CRC Press Web site
http://www.crcpress.com

List of reviewers

Prof. Yasuo **Arai**, President of Japan Institute of Navigation, Japan,
Prof. Eugen **Barsan**, Master Mariner, Constanta Maritime University, Romania,
Prof. Tor Einar **Berg**, Norwegian Marine Technology Research Institute, Trondheim, Norway,
Prof. Carmine Giuseppe **Biancardi**, The University of Naples „Parthenope", Naples, Italy,
Prof. Jarosław **Bosy**, Wroclaw University of Environmental and Life Sciences, Wroclaw, Poland,
Sr. Jesus **Carbajosa** Menendez, President of Spanish Institute of Navigation, Spain,
Prof. Jerzy **Czajkowski**, Gdynia Maritime University, Poland,
Prof. German **de Melo Rodrigues**, Technical University of Catalonia, Barcelona, Spain,
Prof. Eamonn **Doyle**, National Maritime College of Ireland, Cork Institute of Technology, Cork, Ireland,
Prof. Wiliam **Eisenhardt**, President of the California Maritime Academy, Vallejo, USA,
Prof. Wlodzimierz **Filipowicz**, Master Mariner, Gdynia Maritime University, Poland,
Prof. Börje **Forssell**, Norwegian University of Science and Technology, Trondheim, Norway,
Prof. Jerzy **Gazdzicki**, President of the Polish Association for Spatial Information, Warsaw, Poland,
Prof. Witold **Gierusz**, Gdynia Maritime University, Poland,
Prof. Andrzej **Grzelakowski**, Gdynia Maritime University, Poland,
Prof. Lucjan **Gucma**, Maritime University of Szczecin, Poland,
Prof. Jerzy **Hajduk**, Master Mariner, Maritime University of Szczecin, Poland,
Prof. Qinyou **Hu**, Shanghai Maritime University, China,
Prof. Jacek **Januszewski**, Gdynia Maritime University, Poland,
Prof. Bogdan **Jaremin**, Interdepartmental Institute of Maritime and Tropical Medicine in Gdynia,,
Prof. Tae-Gweon **Jeong**, Master Mariner, Secretary General, Korean Institute of Navigation and Port Research,
Prof. Piotr **Jędrzejowicz**, Gdynia Maritime University, Poland,
Prof. Yongxing **Jin**, Shanghai Maritime University, China,
Prof. Nobuyoshi **Kouguchi**, Kobe University, Japan,
Prof. Eugeniusz **Kozaczka**, Polish Acoustical Society, Gdansk University of Technology, Poland,
Prof. Andrzej **Krolikowski**, Master Mariner, Maritime Office in Gdynia, Poland,
Dr. Dariusz **Lapucha**, Fugro Fugro Chance Inc., Lafayette, Louisiana, United States,
Prof. David **Last**, FIET, FRIN, Royal Institute of Navigation, United Kingdom,
Prof. Joong Woo **Lee**, Korean Institute of Navigation and Port Research, Pusan, Korea,
Prof. Józef **Lisowski**, Gdynia Maritime University, Poland,
Prof. Aleksey **Marchenko**, University Centre in Svalbard, Norway,
Prof. Francesc Xavier **Martinez de Oses**, Polytechnical University of Catalonia, Barcelona, Spain,
Prof. Janusz **Mindykowski**, Gdynia Maritime University, Poland,
Prof. Torgeir **Moan**, Norwegian University of Science and Technology, Trondheim, Norway,
Prof. Reinhard **Mueller**, Master Mariner, Chairman of the DGON Maritime Commission, Germany,
Prof. Nikitas **Nikitakos**, University of the Aegean, Greece,
Prof. Stanisław **Oszczak**, FRIN, University of Warmia and Mazury in Olsztyn, Poland,
Mr. David **Patraiko**, MBA, FNI, The Nautical Institute, UK,
Prof. Zbigniew **Pietrzykowski**, Maritime University of Szczecin, Poland,
Prof. Francisco **Piniella**, University of Cadiz, Spain,
Prof. Jens-Uwe **Schroeder**, Master Mariner, World Maritime University, Malmoe, Sweden,
Prof. Chaojian **Shi**, Shanghai Maritime University, China,
Prof. Roman **Smierzchalski**, Gdańsk University of Technology, Poland,
Prof. Henryk **Sniegocki**, Master Mariner, MNI, Gdynia Maritime University, Poland,
Prof. Marek **Szymoński**, Master Mariner, Polish Naval Academy, Gdynia, Poland,
Prof. Lysandros **Tsoulos**, National Technical University of Athens, Greece,
Prof. Dang Van **Uy**, President of Vietnam Maritime University, Haiphong, Vietnam,
Prof. František **Vejražka**, FRIN, Czech Institute of Navigation, „Czech Technical University in Prague, Czech,
Prof. George Yesu **Vedha** Victor, International Seaport Dredging Limited, Chennai, India,
Prof. Peter **Voersmann**, President of German Institute of Navigation DGON, Deutsche Gesellschaft für Ortung und Navigation, Germany,
Prof. Adam **Weintrit**, Master Mariner, FRIN, FNI, Gdynia Maritime University, Poland,
Prof. Adam **Wolski**, Master Mariner, MNI, Maritime University of Szczecin, Poland,
Prof. Jia-Jang **Wu**, National Kaohsiung Marine University, Kaohsiung, Taiwan (ROC),
Prof. Homayoun **Yousefi**, MNI, Chabahar Maritime University, Iran,
Prof. Wu **Zhaolin**, Dalian Maritime University, China

Contents

International Recent Issues about ECDIS, e-Navigation & Safety at Sea. Introduction

A. Weintrit

Gdynia Maritime University, Gdynia, Poland

INTRODUCTION: In the publication there are described international recent issues about ECDIS (Electronic Chart Display & Information Systems) and e-Navigation concept.

1 COUNTDOWN TO ECDIS TIMETABLE

A few years ago the IMO (International Maritime Organization) has established a clear vision for e-navigation, which outlines the direction that shipping and marine navigation communities need to follow.

As we already know, 2009 brought confirmation of the timetable for the mandatory adoption of ECDIS and this means that for large sectors of the industry going digital is no longer an interesting option, it's a must, with an immovable timetable attached.

As a result, companies affected by the first phase adoption in 2012 need to start their planning now.

There are a number of steps and considerations to be made to ensure that there is a smooth transition from paper to digital navigation. The most important thing is finding out how each fleet will be affected – although the legislation will eventually apply to almost all large merchant vessels and passenger ships, it will be phased by vessel type and size so it is vital to know when your ships will be affected. The first phase affects new passenger ships and new tankers.

Developing an implementation strategy is key, as it is important to recognise that the transition from paper to electronic navigation is a fundamental change in the way ship navigation will be conducted, it's not simply a case of fitting another piece of hardware to ensure compliance with a carriage requirement.

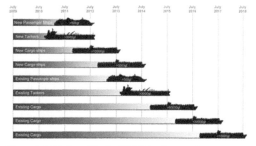

Figure 1. IMO timetable for ECDIS implementation

Key things to consider include the purchase and installation of ECDIS equipment, amendments to bridge procedures, co-ordination between ship and shore, and the selection of a chart service that best meets operational needs and fulfils the carriage requirements.

One of the most important elements is training. Arranging and acquiring the appropriate training certification can take several months and as a minimum you should be able to satisfy your Flag State and any independent audit authorities that your crews are proficient in using ECDIS to maintain safety of navigation.

Although the main aim of ECDIS is safety it can also increase operational efficiency that in turn can lead to bottom-line savings. Navigators and marine superintendents regularly report a steady flow of benefits from using ECDIS, including the fact that updates to chart data can be virtually instant. For more information on how you can successfully adopt and get the best of ECDIS you can read our 10 Steps to ECDIS Mandation by clicking here.

2 ECDIS TRAINING

Electronic Chart Display & Information Systems (ECDIS), improvements in ship to shore communication and what is now global official electronic chart coverage provide mariners with much faster access to the information they need to deliver greater visibility of the marine environment. Ultimately, electronic navigation helps improve the safety of life at sea, as well as drives operational efficiencies for shipping companies.

One of the critical success factors in delivering these benefits to the maritime community lies in ensuring that mariners are able to confidently and safely use ECDIS. An estimated 500,000 personnel will need to be trained on ECDIS navigation over the next few years, as the rolling timetable of deadlines for the ECDIS Mandate come into force from 2012.

To achieve that, there needs to be clear and consistent set of training guidelines based on international agreement, which ensures everyone in the shipping industry knows they have the right training available for compliance with the Mandate, and that it delivers the appropriate skills to crew.

At present, there is still work to be done. The basic requirements are that crew need to undertake two types of ECDIS training: generic ECDIS training, which explains the main features and principles of ECDIS; and manufacturer or type-specific training, which explains the features native to each model. Currently however, there is a lack of uniformity in the training available, and a lack of clarity over the type of training required by different flag states. The UKHO is also concerned that key differences between navigating with paper charts and their electronic versions, Electronic Navigation Charts (ENCs), are not being adequately addressed.

Generic training uses the IMO standard model and is the minimum training required by most flag states. The IMO is making progress to improve the minimum standards of ECDIS training. Last year it ratified amendments to the International Convention of Standards of Training, Certification and Watchkeeping for Seafarers (STCW). This takes into account the new provision for ECDIS training standards and should go some way to setting down the marker for acceptable standards.

Following generic training, mariners need to know the features of the actual ECDIS model they will use at sea. There are over 50 different ECDIS manufacturers, and while all ECDIS models offer essentially the same information, the way the data is presented and the precise controls and features may vary by manufacturer, in the same way that Mac, PC or Linux-based computers differ in user experience. For this reason, it's vital for mariners to attend a type-specific training course.

Finally, in addition to the ECDIS training, mariners need to feel confident managing the information available to them on the ENC. ENCs can offer much richer data to support safe and efficient navigation, but the mariner must know how to quickly and easily interpret that data. The UKHO is working with a number of organisations and to develop an additional training module, which is aimed at helping mariners get the most from the information presented on the ENC.

Navigation with ECDIS has the potential to make a step change in the safety of life and ships at sea and to drive measurable business efficiency along the way. As the first of the deadlines for the ECDIS Mandate approaches, the maritime community as a whole needs to play a part in making sure that the mariner has the right level of support to make the transition from paper to digital navigation as simple, successful and operationally beneficial as possible.

3 E-NAVIGATION CONCEPT

e-Navigation is a concept developed under the auspices of the UN's International Maritime Organization (IMO) to bring about increased safety and security in commercial shipping through better organization of data on ships and on shore, and better data exchange and communication between the two. The IMO decided to include into the work programme a high priority item on "Development of an e-Navigation strategy".

The aim is to develop a strategic vision for e-Navigation, to integrate existing and new navigational tools, in particular electronic tools, in an all-embracing system that will contribute to enhanced navigational safety (with all the positive repercussions this will have on maritime safety overall and environmental protection) while simultaneously reducing the burden on the navigator. As the basic technology for such an innovative step is already available, the challenge lies in ensuring the availability of all the other components of the system, including electronic navigational charts, and in using it effectively in order to simplify, to the benefit of the mariner, the display of the occasional local navigational environment. E-navigation would thus incorporate new technologies in a structured way and ensure that their use is compliant with the various navigational communication technologies and services that are already available, providing an overarching, accurate, secure and cost-effective system with the potential to provide global coverage for ships of all sizes.

e-Navigation remains as a work in progress as its supporting technology continues its evolutionary incorporation into day-to-day maritime operations. Substantive challenges continue to arise as implementation progresses.

During the period 2012 - 2018 vessels subject to SOLAS will be compulsorily fitted with ECDIS, and yet as many as 60% of those ships are using Electronic Charting Systems (ECS) that must be replaced. The required change to ECDIS will require a significant shift in both mind-set and competence. How eNavigation technology evolves to meet challenges like the example above will depend larger on the input from the stakeholder community: the mariner, the regulator, the policymaker …and the manufacturer.

prehensive data in formats that will be more easily understood and utilised by shore-based operators in support of vessel safety and efficiency.
– An infrastructure designed to enable authorised seamless information transfer onboard ship, between ships, between ship and shore and between shore authorities and other parties with many attendant benefits.

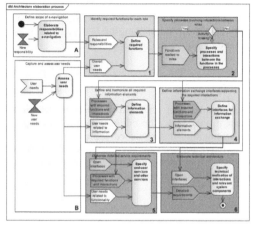

Figure 2. The e-Navigation architecture elaboration process

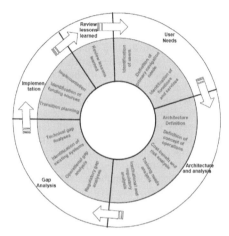

Figure 3. The e-Navigation Architecture

e-Navigation is an International Maritime Organization (IMO) led concept based on the harmonisation of marine navigation systems and supporting shore services driven by user needs.

e-Navigation is currently defined as the harmonised collection, integration, exchange, presentation and analysis of maritime information onboard and ashore by electronic means to enhance berth to berth navigation and related services, for safety and security at sea and protection of the marine environment.

It is envisioned there will be at least three broad significant outcomes from e-Navigation that are currently being used as the basis of establishing user needs. These are represented by ship based systems, shore based systems and a communications infrastructure as outlined here:
– Onboard navigation systems will be developed that benefit from the integration of own ship sensors, supporting information, a standard user interface, and a comprehensive system for managing guard zones and alerts. Core elements of such a system will include high integrity electronic positioning, Electronic Navigational Charts (ENC) and an analysis capability to reduce human error, actively engaging the mariner in the process of navigation while preventing distraction and overburdening.
– The management of vessel traffic and related services from ashore will be enhanced through better provision, coordination, and exchange of com-

4 CONTENTS OF THE MONOGRAPH

The contents of the book are partitioned into five separate parts: e-Navigation concept (covering the chapters 1 through 4), ECDIS - Electronic Chart Display and Information Systems (covering the chapters 5 through 9), visualization and presentation of navigational information, (covering the chapters 10 through 12), data transmission and communica-tion systems (covering the chapters 13 through 19), and safety at sea (covering the chapters 20 through 26).

In each of them readers can find a few chapters. Chapters collected in the first part, titled „e-Navigation Concept", concerning e-Navigation and future trend in navigation, development of requirements for communication management on board in the framework of the e-Navigation concept and advanced maritime technologies to support manoeuvring in case of emergencies as contribution to e-Navigation development. Certainly, this subject may be seen from different perspectives.

In the second part there are described problems related to ECDIS implementation: A harmonized ENC database as a foundation of electronic navigation, navigation safety assessment in the restricted area with the use of ECDIS, integration of the digital selective calling VHF marine radio-communication system and ECDIS to increase the maritime safety, a proposal for a new port related ENC standard (enhance berth to berth navigation requires high quality ENC's – the port ENC), and the overview of the new electronic chart product specification S-101.

Third part is about visualization and presentation of navigational information. The readers can find some information about applications and benefits of cartographic 3D visualisation in maritime safety, assumptions to the selective system of navigational-maneuvering information presentation and security modeling technique: visualizing information of security plans.

The fourth part deals with data transmission and communication systems. The contents of the fourth part are partitioned into seven chapters: maritime communications, navigation and surveillance, data fusion model of the navigation and communication systems of a ship, automation of message interchange process in maritime transport, logical network of data transmission impulses in journal-bearing design, surface reflection and local environmental effects in maritime and other mobile satellite communications, shipborne satellite antenna mount and tracking systems, and yesterday, today and tomorrow of the GMDSS.

The fifth part deals with safety at sea. The contents of the fifth part are partitioned into seven: visual condition at sea for the safety navigation, safety control of maritime traffic near by offshore in time, maritime safety in the Strait of Gibraltar (taxonomy and evolution of emergencies rate in 2000-2004 period), safety at sea – a review of Norwegian activities, improving emergency supply system to ensure port city safety, congested area detection and projection – the user's requirements, and studying probability of ship arrival of Yangshan Port with AIS (Automatic Identification System).

5 CONCLUSIONS

Each chapter was reviewed at least by three independent reviewers. The Editor would like to express his gratitude to distinguished authors and reviewers of chapters for their great contribution for expected success of the publication. He congratulates the authors for their excellent work.

e-Navigation Concept

1. e-Navigation and Future Trend in Navigation

F. Amato, M. Fiorini, S. Gallone & G. Golino
SELEX - Sistemi Integrati, Rome, Italy

ABSTRACT: The International Maritime Organization (IMO) adopted the following definition of e-Navigation: "e-Navigation is the harmonised collection, integration, exchange, presentation and analysis of maritime information onboard and ashore by electronic means to enhance berth to berth navigation and related services, for safety and security at sea and protection of the marine environment". A pre-requisite for the e-Navigation is a robust electronic positioning system, possibly with redundancy. A new radar technology emerged from the last IALA-AISM conference held in Cape Town, March 2010, where almost all the manufactures companies involved on navigation surveillance market presented –at various state of development- solid state products for VTS and Aids to Navigation indicating a new trend for this application. The paper present an overview of the systems for global navigation and new trend for navigation aids. The expected developments in this field will also be briefly presented.

1 INTRODUCTION

The International Maritime Organization (IMO) [1] adopted a "Strategy for the development and implementation of e-Navigation" (MSC85-report, Annexes 20 and 21).

In particular, IMO adopted the following definition of e-Navigation:

"e-Navigation is the harmonized collection, integration, exchange, presentation and analysis of maritime information onboard and ashore by electronic means to enhance berth to berth navigation and related services, for safety and security at sea and protection of the marine environment".

IMO has stated the driving forces and the consequential goal for their e-Navigation concept as follows: << There is a clear and compelling need to equip shipboard users and those ashore responsible for the safety of shipping with modern, proven tools that are optimized for good decision making in order to make maritime navigation and communications more reliable and user friendly. The overall goal is to improve safety of navigation and to reduce errors. However, if current technological advances continue without proper coordination there is a risk that the future development of marine navigation systems will be hampered through a lack of standardization on board and ashore, incompatibility between vessels and an increased and unnecessary level of complexity >> (IMO MSC 85, Annex 20, §2.1).

e-Navigation is therefore a vision for the integration of existing and new navigational tools, in a holistic and systematic manner that will enable the transmission, manipulation and display of navigational information in electronic format.

The paper is organized as follows. After the preparing for e-Navigation in section 2, the initial e-Navigation architecture is presented in section 3 followed by the usage of radar to validate aids to navigation including the identification of buoys by solid state VTS / navigation radars in section 4. Conclusion and future trends are reported in section 5 while references in section 6 will close the paper.

2 PREPARING FOR E-NAVIGATION

IMO has invited IALA and other international organizations to participate in its work and provide relevant input. IALA formed the e-Navigation (e-NAV) committee for the purpose of developing recommendations and guidelines on e-Navigation systems and services. The committee aims to review and develop related IALA documentation on issues such as the impact of new radar technology on radar aids to navigation, future Global Navigation Satellite System (GNSS) and differential GNSS and the impact of electronic ship-borne navigation aids on aids to navigation systems. The committee also works with other international organizations to develop the overall e-navigation concept.

Concerning radars, a clear trend emerged from the last IALA-AISM conference held in Cape Town, March 2010, where almost all the manufactures companies involved on navigation surveillance market present –at various state of development- solid state products for VTS and Aids to Navigation [2][3][4]. The trend of events set the evolution from the microwave tubes (klystrons and magnetron or travelling-wave) [5] to the solid state technologies. It starts from the IMO resolution 192(79) [6] who intend to encourage the development of low power, cost-effective radars removing (from July 2008) the requirement for S-band radar to trigger RACONS (radar beacons). Solid state radar may fulfill these wishes making use of low-power and digital signal processing techniques to mitigate clutter display that are instead associated with high-power magnetron based radars.

A full comparison magnetron versus solid state VTS radars is provided in [7] including experimental result with live data to show clutter filtering and range discrimination of the solid state LYRA 50 radar. The advantage of solid state transmitted could be summarized as long operational life and graceful degradation, coherent processing, high duty cycle, multi frequency transmission on a wide band, no high voltage supply and compact technology.

Regarding GNSS and DGNSS, an overview of the state of art is reported in [8] where also radio aids to navigation are mentioned while an innovative usage of radar to validate aids to navigation (AtoN) is described later in this paper.

In short, based on the IMO definition, three fundamental elements must be in place as pre-requisite for the e-Navigation. These are:

1 worldwide coverage of navigation areas by Electronic Navigation Charts (ENC);
2 a robust and possibly redundant electronic positioning system; and
3 an agreed infrastructure of communications to link ship and shore but also ship and ship.

3 THE INITIAL E-NAVIGATION ARCHITECTURE

In previous sections e-Navigation and its pre-requisite were presented as a concept but in order to implement such concept a technical architecture is needed. It is shown hereafter (Figure 1) where the shipboard entities, the physical link(s) and the shore-based entities are included in this representation.

On the left side is represented, for simplicity's sake, a single "ship technology environment". From the e-Navigation concept's perspective the relevant devices within the ship technology environment are the transceiver station, the data sources and the data sinks connected to the transceiver station, the Integrated Navigation System (INS) and the Integrated

Bridge System (IBS). The transceiver station is shown as a single station for simplicity's sake, although there may be several transceiver stations. The entities which are involved with the specifics of the link technology are confined by the dotted line.

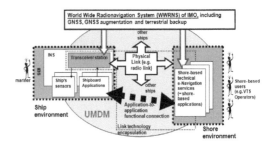

Figure 1. e-Navigation architecture Source: IALA e-NAV140 [9]

The shore-based technical e-Navigation services, in their totality and by their interactions, provide the interfaces of the shore-based user applications to the physical link(s). They also encapsulate their technology to the whole of the common shore-based e-Navigation system architecture. Encapsulation, used as an Object Oriented Programming term is 'the process of compartmentalizing the elements of an abstraction that constitute its structure and behavior; encapsulation serves to separate the contractual interface of an abstraction and its implementation'. The encapsulation principle hides the technology's sophistication from the shore-based e-Navigation system as a whole and thus reduces complexity. The entities which are involved with the specifics of the link technology are confined by the dotted line. Amongst other benefits, it allows for parallel work of the appropriate experts in the particular technology of a given physical link, provided the functional interfaces of the shore-based technical e-Navigation services are well defined.

For the precise technical structure of the shore-base technical e-Navigation services, the common shore-base e-Navigation system architecture is under development for a future IALA Recommendation.

It is also showed the World Wide Radio Navigation System (WWRNS), which includes GNSS, being presented as a system external to the e-Navigation architecture providing position and time information. The Universal Maritime Data Model was also introduced as an abstract representation of the maritime domain [9].

4 USAGE OF RADAR TO VALIDATE AIDS TO NAVIGATION

The use of an additional source of information to improve AtoN has been suggested by Barker in [10]

where he consider the vessel traffic routing information providing by the Automatic Identification System (AIS) as an improvement in the AtoN assessment. On the contrary the on board radar of vessels travelling around an AtoN device could be used to assess and to check the position of the buoys/beacons resulting in an almost real-time verification of the information provided by the Aids and consequently alert for maintenance if needed. To this end, new generation VTS / navigation radars [2] provide a break-through for the task of target classification.

4.1 Identification of buoys by solid state VTS / navigation radars

In this section an example of application of the described system is shown.

Navigation buoys are often affected by drift or malfunctioning. A method to detect the presence of the buoy and the correct position is addressed, and data are sent to ground base station, in order to verify the position and the presence of the buoy against the navigation maps. The method described is used also for maintenance purpose, in case of failure of the device (i.e. low battery) and uses just the passive reflector positioned on the top of the buoy.

Conventional VTS / navigation radars have poor classification capabilities at long distance due to their non-coherent receiver and the limited range resolution. Navigation radars equipped with magnetrons usually are set for medium-high range, to improve safety in navigation, using medium-long pulses (i.e. 30-75 m) with the consequence of very poor discrimination in range.

New generation solid state radars permits to uncouple resolution from transmitted pulse length by using a coherent receiver and long coded pulses. High range resolution can be preserved by implementing a pulse compression algorithm in the digital processor. The radar "LYRA 50" for example can provide a range resolution of 9 m for all ranges up to 24 NM making use of long compressed pulses shown in figure 2.

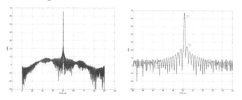

Figure 2. Example of digital a compressed pulse with close up.

The long uncompressed pulse has a time length of 30 μs (4.5 km) and an instantaneous bandwidth of 22 MHz. Due to the digital pulse compression the resulting pulse is severely deformed: it presents one high peak lasting only 60 ns (9 m) and a series of

much weaker peaks (side lobes) lasting all together 60 μs (9 km); however the side lobes are more than 40 dB lower with respect to the main one and their contribute to the received signal is usually negligible.

Range profile can be associated to a radar report and stored for classification purpose by the radar data processor. Buoys can be considered small object for a radar having range extension less than 1 m which produce a narrow peak signal in the radar receiver, whose width equal to the range resolution. Small vessels usually generate a broader peak and sometimes even different discriminated peaks when the length of the vessel is greater than twice the resolution.

An addition contribute to the classification process of the target can be derived by analysis of the amplitude of the echoes. A study on the variation of the amplitude of the target echo over different scans has been conducted by the authors using live radar data (examples are shown see figures 3, 4 and 5).

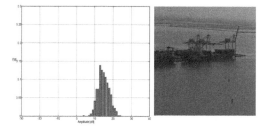

Figure 3. Example of amplitude distribution from an buoy (live data).

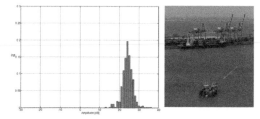

Figure 4. Example of target echoes amplitude distribution from a small anchored boat (live data).

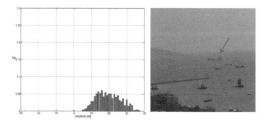

Figure 5. Example of target echoes amplitude distribution from a tanker (live data).

5 CONCLUSION AND FUTURE TREND

For someone in the short term e-Navigation should remain a theoretical concept but it is indubitably an overall concept to which all marine users have to deal with from now on.

Electronic positioning system is a prerequisite for the e-Navigation and position fixing using GNSS is prevailing amongst commercial and leisure users. By the way radars and traditional Aids to Navigation (AtoN) will continue to be required, at least for redundancy and/or terrestrial backup to satellite systems, or in many military scenarios. Co-operation between complementary systems such as radar and AtoN should become stronger and stronger to assessing the data and improving reliability for more informed decisions.

These analysis should be used to improve the filtering and the recognition of buoys, not only based on the expected position data: the recognition process is helpful to distinguish the radar echo of a buoy from other fixed targets such as ships in standby or outcropping rocks.

Further work may include studies on the range extension of the radar echoes.

REFERENCES

[1] Internationa Maritime Office (IMO) website, www.imo.org
[2] F Amato, M. Fiorini, S. Gallone, G. Golino, "New Solid State frontier on radar technologies", Proc. International IALA-AISM Conference 2010, Cape Town, South Africa, 21-27 March 2010
[3] J C Pedersen "A Next Generation Solid State, Fully Coherent, Frequency Diversity and Time Diversity Radar with Software Defined Functionality", Proc. International IALA-AISM Conference 2010, Cape Town, South Africa, 21-27 March 2010
[4] N Ward, M Bransby "New Technology Radars and the Future of Racons", Proc. International IALA-AISM Conference 2010, Cape Town, South Africa, 21-27 March 2010
[5] Collin R.E. "Foundations for Microwave Engineering" 2nd Ed., McGraw-Hill, 1992
[6] IMO MSC79 resolution 192(79) (IMO, 2004)
[7] F Amato, M. Fiorini, S. Gallone, G. Golino, "Fully Solid State Radar for Vessel Traffic Services", Proc. International Radar Symposiums, IRS 2010, Vilnius, Lithuania, 16-18 June 2010
[8] M. Fiorini, "e-Navigation: a Systems Engineering Approach", Proc. MAST 2010, 5th Maritime Systems and Technology conference, palazzo dei congressi, Rome, Italy, 9-11, November 2010
[9] e-NAV 140 "The e-Navigation Architecture – the initial Shore-based Perspective" Ed.1.0, IALA Recommendation, December 2009
[10] R Barker "AIS Traffic Analysis / The Risk Assessment Process for Aids to Navigation", Proc. International IALA-AISM Conference 2010, Cape Town, South Africa, 21-27 March 2010

2. Development of Requirements for Communication Management on Board in the Framework of the E-navigation Concept

F. Motz, E. Dalinger, S. Höckel, & C. Mann
Fraunhofer Research Institute for Communication, Information Processing and Ergonomics, FKIE, Wachtberg, Germany

ABSTRACT: The current separation of communication systems and navigational systems on the ships bridge doesn't meet the requirements of the e-navigation concept of the International Maritime Organization (IMO) for safe navigation to include all means and information in the decision making. Hydrographical, meteorological and safety related information is presented on the communication equipment without filtering or as print-outs solely. A task oriented integration and presentation of this information on the navigational displays will support the officers in their decision making and enhance the safety of navigation. The core element onboard for the integration is the INS (Integrated Navigation System) concept of the IMO where a task and situation dependent presentation of information is specified based on a modular concept. Information should be automatically processed, filtered and integrated in the navigational information systems to support the users in their tasks. To achieve this goal a concept for communication management was developed. An Applied Cognitive Work Analysis (ACWA) is conducted to identify requirements for the design of a communication management system based on the cognitive processes of the operators. This paper describes the concept for communication management and, as a first result, gives the description of the domain of maritime communication that provides a basis for the identification of requirements for communication management in the framework of the e-navigation concept.

1 INTRODUCTION

The International Maritime Organization (IMO) identified the need to equip shipboard users and those ashore responsible for the safety of shipping with modern, proven tools optimized for good decision making in order to make maritime navigation more reliable and user friendly. In this framework the IMO decided on proposal of several member states to develop an e-navigation strategy to integrate and utilize new technologies in a holistic and systematic manner to make them compliant with the various navigational, communication technologies and services that are already available.

The e-navigation strategy aims to enhance berth to berth navigation and related services by harmonizing the collection, integration, exchange, presentation and analysis of marine information onboard and ashore by electronic means (IMO, 2007a). The e-navigation strategy is supposed to be user-driven not technology driven to meet present and future user needs (IMO, 2007a).

On behalf of the German Ministry of Transport, Building and Urban Affairs (BMVBS) an e-navigation user needs survey was conducted (Höckel & Motz, 2010; IMO, 2009a; IMO, 2009b). One of the major issues which were identified was the need for user-selectable presentation of information received via communication systems on the navigational displays of the ships bridge. This need relates to, e.g., hydrographical, meteorological and safety related information, and was also found in user needs assessments of other member states and organizations provided to the IMO (IMO, 2009c).

The integration and presentation of information on board pertaining to planning and execution of voyages, assessment of navigational risk and compliance with regulations is an essential part of the e-navigation concept. The current separation of communication systems and navigational systems doesn't meet the e-navigation requirements of safe navigation to include all means and information in the decision making. Hydrographical, meteorological and safety related information is presented on the communication equipment without filtering or as print-outs solely. Technical as well as legal conditions (separation of responsibilities in the Safety of Life at Sea Convention – SOLAS chapters IV and V) hamper the integration of information provided by communication equipment in the navigational systems, which reduces their utilization.

A task-oriented integration and presentation of this information on the navigational displays considering the fact that all necessary information for the respective task and situation, is on its disposal fast, reliable, consistent and easily interpretable will support the officers in their decision making and enhance the safety of navigation. The task-oriented approach for presentation and integration of navigational information as introduced with the revised performance standards for Integrated Navigation Systems (INS) (IMO, 2007b) based on the modular concept (IMO, 2008a) forms the basis for the integration of further information on board. Modular Integrated Navigation Systems (INS) according to the revised IMO performance standards for INS (IMO, 2007b) combine and integrate the validated information of different sensors and functions and allow the presentation on the various displays according to the tasks.

A communication management system should be employed on the bridge as an aid for the mariner in the accomplishment of communication tasks and as a mean for the provision of information to INS.

2 INS AND MODULAR BRIDGE

The aim of the INS specified according to the IMO performance standards (IMO, 2007b) is to promote safe procedures for the integration of navigational information and to allow that an INS is considered as "one system" that is installed and used as other means under SOLAS Chapter V regulation 19, instead of stand-alone navigational equipment onboard ships. These performance standards can be applied via a modular concept for either comprehensive integrations which are specified as INS or only partial integrations.

The purpose of an INS is to enhance the safety of navigation by providing integrated and augmented functions to avoid geographic, traffic and environmental hazards. The INS aims to be demonstrably suitable for the user for a given task in a particular context of use. An INS comprises navigational tasks such as "Route planning", "Route monitoring", "Collision avoidance", etc. including the respective sources, data and displays which are integrated into one navigation system. An INS is defined as such in the performance standards, if it covers at least two of the following navigational tasks / functions:
- Route monitoring
- Collision avoidance

An alert management is a required functionality of the INS as well as the presentation of navigation control data for manual control. Other navigational tasks may also be integrated into the INS.

The following six navigational tasks are described in detail within the performance standards for INS (IMO, 2007b):

- Route planning
- Route monitoring
- Collision avoidance
- Navigation control data
- Alert management
- Status and data display

The scope of the INS may differ dependent on the number and kind of tasks and functions integrated into the INS. The performance standards for INS (IMO, 2007b) allow for a differentiated application of the requirements depending on integrated task and functionality.

With regard to the integration of information received via communication systems on the navigational displays the INS performance standards allow for the provision of tidal and current data, weather data, ice data, and additional data of the tasks 'navigation control' and 'route monitoring' on the status and data display. For 'route planning' the INS provides means for drafting and refining the route plan against meteorological information if available in the INS, while for navigational purposes, the display of other route-related information (e.g., monitoring of SAR manoeuvres, NAVTEX, weather data, etc.) on the chart display for 'route monitoring' is permitted.

With the modular bridge concept (Fig. 1), operational/functional and sensor/source modules are specified. This allows clear separation between operational requirements for the task orientated use and presentation of information on equipment and systems, and between the sensor specific technical requirements. The interfacing module specifies the connection and data exchange with other systems. Based on the modular bridge concept the design of future systems becomes flexible, task and situation orientated.

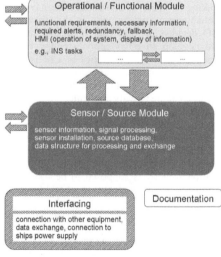

Figure 1. Modular Bridge Concept

3 CONCEPT FOR INTEGRATION OF INFORMATION RECEIVED VIA COMMUNICATION SYSTEMS

For developing a concept for communication management for the ships bridge the current communication infrastructure and procedures on board were analyzed. A literature review was conducted with regard to GMDSS required systems and additional technology. To identify further aspects for the determination of user requirements regarding communication management and conditions regarding the transfer of information from the communication systems into the navigational systems interviews with potential users were carried out.

Based on this information a concept for communication management was developed with the following objectives:
- Presentation of information received via communication systems on the navigational displays (INS) of the ships bridge
- User-selectable automatic filtering and processing of information to prevent information overload
- Provision of source and channel management (selection of best connection according to criteria, e.g., content, integrity, costs)
- Increased availability and reliability due to efficient use of different communication channels

Figure 2 provides a content-related visualisation of the communication management concept.

Basically an INS task for "Communication Management" is introduced to cluster information from different communication systems according to information type. Information is then routed to the navigational and other bridge systems or may be provided on request. Information clusters reflect the information types identified in the analysis of the communication infrastructure:
- Emergency messaging
- Navigational information
- Meteorological information
- Hydrographical information
- Reporting
- Communication with office
- Crew and passenger communication
- Special purpose applications

While data acquisition and data communication remain with the established communication systems, the communication management system provides source and channel management. This means that the connection for data communication may be selected based on certain criteria, e.g., integrity, content and costs. Criteria are adjusted at the human machine interface of the communication management.

Within the communication management system data is evaluated and clustered according to information type. Further processing and filtering allows for updating of previously received information,

avoidance of information doubling, and selection of information relevant to the vessels' type and route.

Data is stored in a database to be provided to other, e.g., navigational, systems on request, but also for presentation on the user interface of the communication management system. This presentation gives an overview of the data according to information type, data source, and time of reception. The user interface also serves as input device for the setting of parameters for filtering as well as source and channel management.

This concept for communication management provides the functionality for the presentation of information received via communication systems on the navigational displays of the ships bridge. The basis of this integration, however, lies in the modular, task-oriented bridge design.

The introduced communication management functionality can be integrated in the task-orientated concept described within the INS performance standards (IMO, 2007b) and the modular bridge concept (Motz et al., 2009b). It can be specified as a new INS task or as part of the INS task "status and data display" to allow for the management and routing of the information received via communication systems into the bridge systems for presentation and use.

The concept, nevertheless, is provisional and needs to be further investigated. In the following an approach for designing a communication management system is introduced, which examines the human cognitive processes for decision making. In order to develop requirements for the graphical user interface of the communication management system on board an Applied Cognitive Work Analysis (ACWA) is conducted.

4 METHOD FOR DETERMINATION OF REQUIREMENTS FOR COMMUNICATION MANAGEMENT

4.1 Problem description

Ship's bridge systems are often designed by aggregation of different computer-based systems, which are developed independent from each other by different suppliers. In this manner new technologies are developed and integrated separately and the displays are designed according to the equipment, which is behind it, and not according to the cognitive demands of the task. The system development based on new technology is a bottom-up approach, where smaller subsystems are defined before linking them together to a large top-level system. To develop interfaces, which support efficient decision making of nautical officers, a top-down approach is required with design methods, which examine human cognitive processes and determine human cognitive needs.

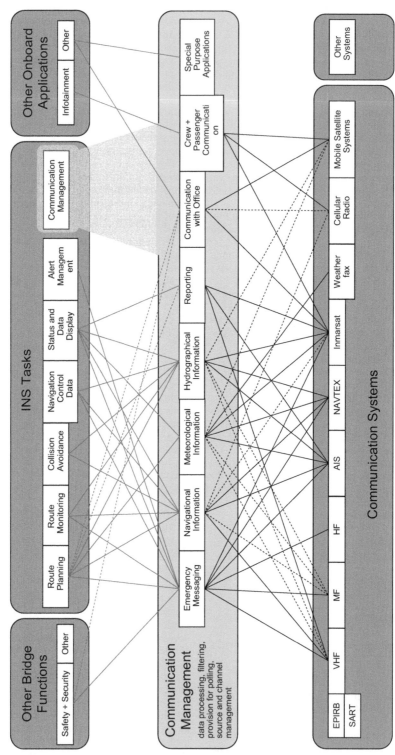

Figure 2. Concept for the integration and management of information received via communication systems

Supporting the decision-making process of nautical officers demands understanding how decisions are made in real-world situations. The design of the navigation systems should take into account the cognitive demands of nautical officers to support them in their work.

A communication management system must not only manage the transfer of information, but also the integration of information, which are received via the communication systems, as this information must be processed and forwarded to a relevant task station in real-time to guarantee quick response.

4.2 *Applied Cognitive Work Analysis (ACWA)*

Cognitive Systems Engineering (CSE) is a design framework, which focuses on analysis of cognitive demands in order to identify cognitive processes of operators. For understanding cognitive demands of the people it is necessary to understand the environment in which people are acting. The environment can be the physical properties of a workplace, the demands of the tasks, the structural characteristics of the work domain or the organizational structure of the company. CSE is primarily focused on the work domain, its constraints, and goals to be reached in the domain. Methods of CSE help to understand, how experts make decisions and why they make certain decisions, what cues they need, what knowledge and strategies they use. Applied Cognitive Work Analysis (ACWA) is a method of CSE for the analysis, design and evaluation of complex systems and interfaces. ACWA applies the Rasmussen's abstraction hierarchy (Rasmussen, 1985) which describes the human information processing. With the ascending in the hierarchy the understanding for goals to achieve rises. Moving to deeper levels yields better understanding for the system's functions with a view to achievement of these goals.

ACWA comprises the following process steps (Elm at al., 2003):
- Development of the Functional Abstraction Network (FAN) – a model to represent the functional relationships between the work domain elements. Each node in the network represents a goal, links represent support.
- Identification of cognitive demands which arise in the domain and need support – Cognitive Work Requirements (CWR) or decision requirements. At this step decisions, which are to be made to achieve the goals defined in the FAN, should be identified.
- Identification of the Information / Relationship Requirements (IRR) for effective decision-making. At this step the information required for each decision should be defined, which is of particular importance, given that the decision making

is based upon the interpretation of information. Incorrect or incomplete information leads to wrong decisions.
- Definition of a relationship between the decision requirements and visualization concepts – Representation Design Requirements (RDR). This step defines how the information should be represented. The decision-aiding concepts should be developed on the basis of information requirements taking into account human perception and cognition.
- Implementation of representation requirements into a powerful visualization of the domain context – Presentation Design Concepts (PDC). A prototype, which supports the cognitive tasks through appropriate visualization, should be developed.

The ACWA depicts an iterative process, as with the development of a prototype to evaluate the effectiveness of the new system additional cognitive and information requirements for decision support, which were missed in the first steps, could be identified.

5 WORK DOMAIN ANALYSIS OF MARITIME COMMUNICATION

The first step of ACWA (building of FAN) was applied to the domain of maritime communication. Diverse knowledge elicitation techniques, such as reviewing relevant documents and interviews, were used to gain understanding of the domain. The purpose of FAN is to provide a base for definition of design requirements for a user interface for the communication management on board.

In Figure 3 the three main functional areas of the communication management are presented. One of the functions of the communication management is the exchange of information with the outside world, e.g., shore authorities, other ships (3). Another function is the exchange of information with own ship systems such as INS tasks (1). The exchange of information comprises the forwarding of information received through the communication equipment to systems on the bridge or INS tasks and the request of information from systems or INS tasks, which is intended for transmission via communication systems. Another function is the information management, which comprises retrieval and processing of both external and internal information (2).

Figure 4 represents the FAN for the domain of maritime communication. The goals are marked according to corresponding functional areas of the communication management system (see Fig. 3).

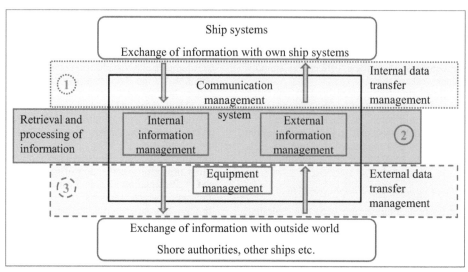

Figure 3. Three main functional areas of the communication management.

First of all the purposes and external constraints in the work domain were identified. The high-level goals are:
– Ensuring safety and security
– Attending administrative matters (organizational norms and goals)
– Attaining commercial goals (achieving the optimum turnover)
It is important to keep these high-level goals in mind while specifying the goals on the lower levels because of the norms and constraints that they define. To ensure these goals it is necessary to
– Navigate the ship safely
– Navigate the ship efficiently
– Keep the ship in an operable state
One of the most important goals, which supports all of the goals mentioned above, is the faultless management and maintenance of communication and information processing (G1). Communication plays an important part in gaining and forwarding information. The ship needs to provide information to diverse authorities on shore. The shore authorities, e.g., Vessel Traffic Services (VTS), shipping companies or port facilities, need to communicate important information to ships. Furthermore, the ship needs to receive (updates of) navigational, hydrological, meteorological and other information on a regular basis.

To successfully manage the communication and information processing (see Fig.4) the management of transmission and reception of information and also forwarding of the received information to eligible ship systems are required. To successfully manage the information transmission (G2) first of all it is necessary to summon the relevant information,

which is requested by others or is scheduled to be transmitted (e.g., as a report). After the information is sent the storage of transmitted data takes place in order to provide a proper documentation of communication activities.

To manage the information reception (G3) it is important to control the incoming information in order to enable secure and faultless data transfer, which comprises recovery, decoding, and verification of received data. All information received via communication systems must be identified, evaluated and stored before it can be forwarded to other systems or used in any other manner.

To successfully manage the internal information forwarding (G4) the observation of incoming information must take place. The actual information, which is necessary to fulfill navigational or other tasks must be forwarded to eligible on-board systems and information requests from those systems must be processed.

Monitoring of the received information (G6) enables detection of new information or changes, so that notifications of the availability of new information can be made or existing information can be updated. It is essential to enable secure and error-free data transfer and provide the possibility to restore messages in case of transfer failures. Reliability control of information (G7) is, therefore, important to ensure security and safety of ship operations. On the other hand the classification of information must take place in order to identify the belonging of information to a certain information type.

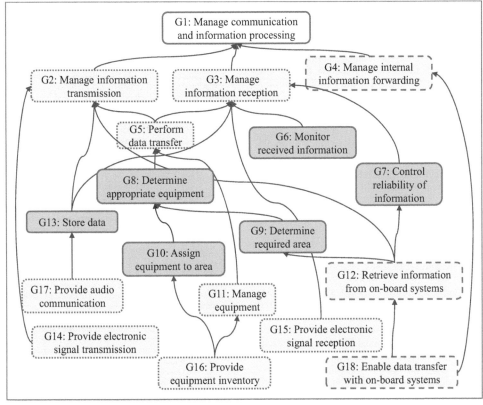

Figure 4. The FAN for the domain of maritime communication.

Additionally, plausibility checks should be made with information sources onboard and between information provided by external domains to avoid possible errors and inconsistencies, which information from different sources could contain.

In order to transmit or receive information it is necessary to successfully perform data transfer (G5), which includes providing of appropriate equipment for transfer and monitoring of data transfer in order to determine failures. The important step in the performance of data transfer is the determination of appropriate equipment (G8) for reception or transmission of information. This includes the determination of communication area (G9) and assignment of equipment to area (G10), which is important for planning the data transfer. First of all the area, where the ship operates, must be identified. Before choosing the appropriate equipment it is essential to determine, which communication equipment necessary to fulfill a certain communication task is available and/or adequate for the task. And that means to check, whether the available equipment is optimal, suitable or inappropriate in respect to its content, required bandwidth, costs... The equipment management (G11) comprises checking of functions, configuration

(channel management) or troubleshooting of equipment. To enable successful equipment management a proper equipment inventory (G16) must exist. Location, connections, description and state of the communication equipment must be provided to support the decision maker in taking required measures, e.g., in case of failures.

Further, transferred data must be stored (G13) and communication log must be generated. The information, which is received via communication systems, contains, e.g., navigational, meteorological, hydrographical, geographical and voyage data. Arrangement of data according to the type of communication equipment, the information type or the transmitting station must be provided.

Information retrieval from on-board systems (G12), which comprises the sending of a data request to other own ship systems and the reception of data from them, is necessary to provide intern ship information in order to transmit it via communication equipment, to compare incoming information with intern ship data for verification purposes, and to determine the communication area. The supporting goal (G18) is to enable the data exchange with on-board systems.

At last the supporting goals at the lower level, such as reception and transmission of audio and digital maritime communication information, are to be mentioned: the provision (G14) and reception (G15) of electronic signal transmission, as well as the provision of audio communication (G17). The latter is necessary to allow for the possibility to log audio communication information.

6 CONTINUATION

The FAN provides a basis for the definition of requirements for a user interface for the communication management on board. In further ACWA steps for each goal of the FAN the decision requirements, information requirements and decision-aiding concepts will be identified.

The relationship between the goals in the domain, the cognitive demands of nautical officers and the information required to make decisions are factors, which provide a basis for designing visual aids for decision support. The further steps are the development of a prototype for the graphical user interface of the communication management system on board and its evaluation.

The concept for communication management will be evaluated in expert reviews, e.g., with members of the national e-navigation working group. Interviews and observations will be conducted onboard ships to gain further insight into the circumstances and challenges of communication during usual operating procedure, and what kind of information is required for which INS task.

First solutions for the design of the human machine interface of the communication management will be developed as paper prototypes and evaluated in brief user tests.

ACKNOWLEDGEMENTS

The research is part of a project funded by the German Ministry of Transport, Building, and Urban Affairs.

REFERENCES

Burns C. M. & Hajdukiewicz J. R. 2004. Ecological Interface Design. Boca Raton, FL: CRC Press.

Dalinger, E. & Motz, F. 2010. Designing a decision support system for maritime security incident response. In O. Turan, J. Boss, J. Stark, J.L. Colwell (eds). Proceedings of International Conference on Human Performance at Sea.

Elm, W., Potter, S., Gualtieri, J., Roth, E. and Easter, J. 2003. Applied cognitive work analysis: a pragmatic methodology for designing revolutionary cognitive affordances. E. Hollnagel (ed.), Handbook for Cognitive Task Design, London: Lawrence Erlbaum Associates.

Höckel, S. & Motz, F. 2010. Determination of User Needs for Future Shipboard Systems in the Framework of the IMO e-Navigation Strategy. In: Turan, O.; Bos, J.; Stark, J.; Colwell, J.L. (Eds.): International Conference on Human Performance at Sea Proceedings: 535-544. University of Strathclyde, Glasgow, United Kingdom.

IMO 2007a. Report of the Subcommittee on Safety of Navigation to the Maritime Safety Committee. NAV 53/22. London: International Maritime Organization.

IMO 2007b. Revised Performance Standards for Integrated Navigation Systems (INS) MSC.252(83). London: International Maritime Organization.

IMO 2008a. Guidelines for the application of the modular concept to performance standards. SN.1/Circ.274. London: International Maritime Organization.

IMO 2009a. Development of an e-navigation strategy implementation plan - Results of a worldwide e-navigation user needs survey. Submitted by Germany to the 55th session of the IMO Sub-Committee on Safety of Navigation. NAV 55/11/3. London: International Maritime Organization.

IMO 2009b. Development of an e-navigation strategy implementation plan - Results of a worldwide e-navigation user needs survey. Submitted by Germany to the 55th session of the IMO Sub-Committee on Safety of Navigation. NAV 55/INF.9. London: International Maritime Organization.

IMO 2009c. Development of an e-navigation strategy implementation plan - Mariner needs for e-navigation. Submitted by the International Federation of Shipmasters' Associations (IFSMA) to the 55th session of the IMO Sub-Committee on Safety of Navigation. NAV 55/INF.8. London: International Maritime Organization.

Rasmussen, J. 1985. The Role of Hierarchical Knowledge Representation in Decisionmaking and System Management. IEEE Transactions on Systems, Man, and Cybernetics. SMC-15 (2): 234-243.

Vicente, K.J. & Rasmussen, J. 1992: Interface Design: Theoretical Foundations. IEEE Transactions on Systems, Man, and Cybernetics, 22 (4): pp. 589-606.

e-Navigation Concept

International Recent Issues about ECDIS, e-Navigation and Safety at Sea – Marine Navigation and Safety of Sea Transportation – Weintrit (ed.)

3. Advanced Maritime Technologies to Support Manoeuvring in Case of Emergencies - a Contribution to E-navigation Development

M. Baldauf, S. Klaes & J.-U. Schröder-Hinrichs
World Maritime University, Maritime Risk and System Safety (MaRiSa) Research Group, Malmoe, Sweden

K. Benedict & S. Fischer
Hochschule Wismar, Dept. of Maritime Studies, ISSIMS, Rostock-Warnemuende, Germany

E. Wilske
SSPA Sweden AB, Gothenburg

ABSTRACT: Safe ship handling in every situation and under all prevailing circumstances of ship status and the environment is a core element contributing to the safety of the maritime transportation system. Especially in case of emergencies, there is a need for quick and reliable information to safely manoeuvre a ship e.g. to quickly return to the position of a Person-overboard (PoB) accident. Within this paper investigations into onboard manoeuvring support for Person-overboard accidents will be presented. Based on the analysis of selected accident case studies and existing solutions representing the technical state-of-the-art, shortcomings will be identified and discussed and a potential approach for advanced manoeuvring support in the context of e-Navigation based requirements will be introduced and discussed.

1 INTRODUCTION

One substantial contribution to safety of the maritime transportation system is safe ship handling. It has to be realised in every situation and under all potential prevailing circumstances of the ship status (i.a. characterised by ship type and shape, draught etc.) and the environment (as, e.g., water depth, wind, current etc.). In the case of certain dangers or concrete emergencies there is an urgent need for quick and reliable information in order to safely manoeuvre a ship e.g. to quickly return to the position of a Person-overboard (PoB) accident. Especially in such situations, manoeuvring information provided by standard wheelhouse posters or the required standard manoeuvring booklet are inconvenient and insufficient.

According to the definitions given by IMO/IALA, e-Navigation is the harmonised collection, integration, exchange, presentation and analysis of maritime information onboard and ashore by electronic means to enhance berth-to-berth navigation and related services, for safety and security at sea and protection of the marine environment. Within this concept new approaches to provide advanced manoeuvring support in case of emergencies can also be developed.

There are ongoing investigations into potential enhancements for onboard manoeuvring support and assistance for the specific case of Person-overboard accidents. Among others motivated by the introduction of new information and communication technologies and their potentials for more sophisticated solutions, research and development activities taking also into account the latest e-Navigation initiative of IMO and IALA have been started. Based on analysis of selected accident case studies and existing solutions, representing the technical state-of-the-art, lacks and shortcomings will be identified and discussed in the next chapter followed by development of a concept for advanced situation-dependent manoeuvring support. Relations to and requirement derived from IMO's and IALA's e-Navigation initiative will be introduced and discussed.

2 PERSON OVERBOARD CASE STUDIES

2.1 *Container ship in heavy seas*

A fully laden containership was on a voyage from port of Rotterdam sailing to a port in the Baltic Sea. The actual speed was reduced due to deteriorating weather conditions with strong winds and increasing wave heights.

Some bunker room alarms occurred during night time and the ship command decided to send a team to investigate the situation and the source of alarms.

A team of two engineers went to the bunker room between cargo holds but one of them was hit by a wave and washed overboard. Although his immersion suit was without a floating device, it kept him warm and a fender which had been torn loose kept him afloat.

The ship's command immediately informed shore-based traffic centre and requested assistance but decided not to conduct a return manoeuvre such as a SCHARNOW- or WILLIAMSON-Turn and, as documented in the official accident investigation report, continued her voyage without changing course or speed at all.

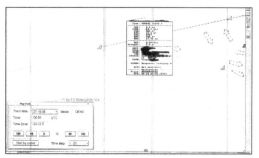

Figure 1. Snapshot from ECDIS record – continuation of the voyage without conduction any manoeuvre (taken from accident investigation report)

The person overboard was several hours later rescued by a SAR vessel and brought to a hospital, where he was recovered and was able to resume his work a few days later.

2.2 *Person-overboard in open sea area*

A container ship was en route from Mexico to Japan in winter. The ship's route had to lead through a sea area behind a hurricane. However, the average wind condition during the time of the accident was Bft 5 with corresponding sea state and with significant wave heights of approximately 5m. In the sea area 300 nm off the Japanese coast a team of four crew members performed various tasks on the bow. In the course of their work several strong waves washed over the deck, hitting three seamen and sweeping overboard one of the mariners.

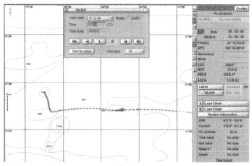

Figure 2. Manoeuvre track during person overboard accident (taken from official accident investigation report)

According to the official accident investigation report, which also includes an analysis of the manoeuvre log and ECDIS records, there were no manoeuvres in compliance with or at least similar to

one of the known return manoeuvres to bring the ship back to the accident's position, or on a opposite course along the original track.

While some crew members attempted resuscitation, others were involved in search measures initiated by the ship command. The resuscitation efforts in connection with the seriously injured mariner were unsuccessful. Darkness started to fall as early as 1700 hours. Although there were supporting search efforts by Japanese Coast Guard (JCG) aircraft, the seaman who had been gone overboard could not be found. In addition at around 2100 hours rain began to fall. The search was ultimately suspended six hours after the accident due to continuously deteriorating weather, and resumed the next day by the JCG.

The vessel finally continued its journey to Japan, where two injured mariners recovered in hospital. The mariner who had been swept overboard was never found.

3 INVESTIGATIONS INTO THE PRESENT SITUATION AND STATE OF THE ART

As demonstrated in the cases studies above, even today a person overboard accident in most cases unfortunately ends with the death of the concerned person. Available statistics from national Marine Accident Investigation branches all over the world show that in up to 75% (see e.g. Annual Marine Incident Report 2003, Queensland) of such cases a mariner or passenger overboard finally died. Several publications refer to an average number of 1,000 dead worldwide per year due to person overboard accidents. According to the latest information about cruise and ferry passengers and crew overboard accidents only of North American passenger shipping companies, compiled by KLEIN for the period from 2000 to 2010, there were over 150 PoB accidents.

Compared to groundings and collisions, person overboard accidents are rarer events but in terms of risk assessment have much greater consequences. A person overboard accident requires immediate decision making and prompt action. Every second is important and influences the success of the actions to rescue the person overboard.

There are standard plans available which can be visualised, e.g., as flow chart diagram as exemplarily shown in the figure below.

But the poor success rate of rescue actions begins already with the difficulties of recognising the event immediately. The first task of the bridge team is to mark the position, release a life ring with safety buoy (smoke and light signal), keep sharp look out and turn the ship back to the position of the accident to pick up the person overboard.

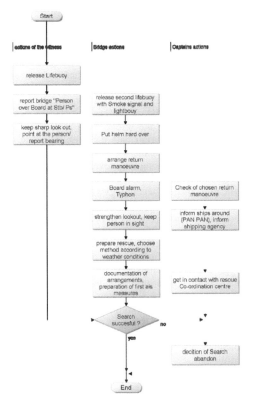

Figure 3. Sample of a Person-overboard action plan addressing actions of witness, captain and members of the bridge team (according to HAHNE)

The crucial action is to bring the ship back to the position of the accident. In literature, several manoeuvres for person-overboard accidents are described. However, there is no single standard procedure recommended, as the effectiveness of a manoeuvre depends on the type of the ship and the prevailing circumstances of the particular situation. The guidance given therefore basically takes into account only the amount of time passed after the accident. According to the IAMSAR (International Aeronautical and Maritime Search and Rescue) / MERSAR (Merchant Ship Search and Rescue) Manual, firstly published by IMO in 1970, threefold action cases for manoeuvring are described:
– "Immediate action" situation,
– "Delayed action" situation and
– "Person missing" situation.

Referring to the experiences and proven effectiveness in many person-overboard casualties the SINGLE-TURN, the WILLIAMSON-Turn as well as the so called SCHARNOW-Turn are mentioned in the MERSAR manual. However, there are further turns which are rarely used in commercial shipping as knowledge and/or experience is limited. In case of real accidents, almost no experience is available for most of the ship officers; they never or seldom have experienced such an accident personally.

The mandatory training procedures, including the conduct of return manoeuvres, are normally executed under good conditions in order to keep a safe environment for persons involved in the training routine. Contrary to this, in accidents the conditions, especially the wind and waves, are worse. Action plans according to the International Safety Management (ISM) Code are available, but in the case of real situations the use of these plans is often limited because plans are made to give more general guidance. No technical means, or only unsuitable ones, are available e.g. for the immediate selection and planning of the manoeuvre in the respective situation.

Today ECDIS and GPS or other systems are available to allow for marking the position of an accident electronically. However, it has to be done manually. As accident investigations have shown in such stressful situation the crew member may fail to do so.

Most Radar/ECDIS equipment available on the market (i.a. Transas NaviSailor or Furuno ECDIS EC 1000) allows the display of a marked position and may provide information about distance and Time To Go (TTG) to the marked position on basis of calculation using actual course/speed information. Some more enhanced systems (e.g. latest Visionmaster FT systems of Sperry Marine) even allow for the display of search patterns – but this is needed later, if the immediate measures for finding the person right after the accident have failed.

The consideration of external factors, such as wind influence on the ship's track, is possible only on the basis of the mental model of the ship officer on watch; no computer-based support is available when it is most urgently needed.

Like all other maritime accidents, person-overboard and search and rescue cases are rare events. Immediate actions are necessary and have to take into account the prevailing circumstances of the environment and the manoeuvring characteristics of the ship. The general guidelines and information for manoeuvring have to be adapted to the actual situation. However, the manoeuvring data displayed on paper on the bridge to assist the captain and navigating officers are of a general character only and of limited use in the case of real accidents. Manoeuvring assistance regarding optimised conduction adapted to the specific hydrodynamic and the actual environmental conditions are urgently needed.

Although new and highly sophisticated equipment and integrated navigation and bridge systems (INS / IBS) have great potential to provide enhanced assistance, situation-dependent manoeuvring information and recommendation are not available yet. The same is true for SAR actions. Optimisation and coordination of all involved parties is needed, taking into account e-Navigation related concepts.

Finally, the related training courses need to be enhanced, especially by means of the use of full-mission ship-handling simulation facilities.

4 INTEGRATED MARITIME TECHNOLOGIES FOR ADVANCED MANOEUVRING ASSISTANCE

4.1 Selected Aspects of Manoeuvring

Ship manoeuvres can be divided into routine manoeuvring and manoeuvring in safety-critical and emergency situations. This division can be developed further by considering different sea areas where manoeuvres have to be performed: e.g. in open seas, in coastal waters and fairways as well as in harbour approaches and basins. Routine manoeuvring in open seas covers ship-handling under normal conditions, e.g. in order to follow a planned route from the port of departure to the port of destination, and include simple course change manoeuvres, speed adaptations according to the voyage plan etc.

Manoeuvring in coastal areas, at entrances to ports and in harbour basins include manoeuvres, e.g. to embark and disembark a pilot, to pass fairways and channels and even berthing manoeuvres with or without tug assistance.

Manoeuvring in safety-critical and emergency situations deals with operational risk management and includes manoeuvres to avoid a collision or a grounding, to avoid dangerous rolling in heavy seas, or to manoeuvre in the case of an real accident e.g. return manoeuvres in case of a person overboard accident or when involved in Search-and-Rescue operations.

Taking the case studies described in the second section it can be concluded that there is a strong need to improve and support the ship command with more sophisticated situation-dependent manoeuvring information, especially in an emergency. It is worthwhile to use the potential of e-Navigation and the related new technology in order to generate such assistance to the human operator when a person has fallen overboard.

4.2 Situation dependent manoeuvring assistance by dynamic wheelhouse poster and electronic manoeuvring booklet

As earlier investigations (Baldauf & Motz, 2006) into the field of collision and grounding avoidance have shown, there is an unsatisfactory exchange of information which is already available on a ship's

navigational bridge from different sensors and sources.

Until today the change of manoeuvring characteristics, e.g. with respect to their dependencies on speed and loading conditions, as well as on environmental conditions (e.g. water depth, wind and current) has not yet been sufficiently considered. High sophisticated Integrated navigation systems (INS – see also IMO, 2009) are installed on board but do not provide the bridge team with situation-dependent manoeuvring data e.g. turning circle diameter, stopping distances etc. for the actual situation. However, the ongoing developments under the IMO's and IALA's e-Navigation initiative with the application of new technologies and data might allow exactly this in the future. In the context of the e-Navigation concept and its definition, the introduction of a dynamic wheelhouse poster and an electronic manoeuvring booklet are suggested. Up-to-date manoeuvring information adapted to specific purposes and situations can be provided by using enhanced integrated simulation technologies.

For that purpose a first generic concept has been drafted to combine own ship status and environmental information from different sensors and manoeuvring information that, e.g., could be gained via a mandatory Voyage Data Recorder or from ECDIS recordings.

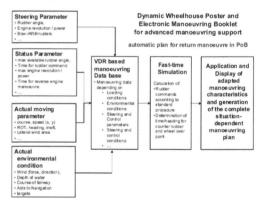

Figure 4. Principal structure and data-flow for generating a dynamic wheelhouse poster and manoeuvring booklet to provide situation dependent manoeuvring support for return manoeuvre

For a person overboard accident the mandatory wheelhouse poster should contain information about return manoeuvres. Spotlight analyses have shown that in most cases this information is incomplete and only partly or not available in the documents, even for the basic cases of deep and shallow waters as well as for loaded and ballast conditions.

4.3 Application of fast-time simulation techniques for Manoeuvring Assistance

The following equation of motion is used as the model for the ships dynamic and implemented in software modules for fast time simulation:

$$X = m(\dot{u} - rv - x_G r^2)$$
$$Y = m(\dot{v} + ru + x_G \dot{r}) \qquad (1)$$
$$N = I_z \dot{r} + m x_G (\dot{v} + ru)$$

On the right side are the effects of inertia where u and v represent the speed components in longitudinal and transverse direction x and y, and r is the rate of turn of the ship. The ship's mass is m, and x_G is the distance of the centre of gravity from the origin of the coordinate system, I_z is the moment of inertia around the z-axis. The ship's hull forces X and Y as well as the yawing moment N around the z-axis are on the left side. Their dimensionless coefficients are normally represented by polynomials based on dimensionless parameters, for instance in the equation for transverse force Y and yaw moment N given as the sum of terms with linear components N_r, N_v, Y_r and Y_v and additional non-linear terms. Other forces, such as rudder forces and wind forces are expressed as look-up tables. There are additional equations for the engine model, and also look-up tables to represent automation systems characteristics. The solution of this set of differential equations is calculated every second; some internal calculations are even done at a higher frequency. Further detailed descriptions can be found in Benedict (2010).

The inputs for the simulation module consist of controls, the states and the data for the environmental conditions. Additionally, there is an input of the ship's condition parameters. They are normally fixed but in case of malfunctions they might change, e.g. reducing the rudder turning rate or maximum angle. The results from the simulation module are transferred to be stored or directly displayed on demand in the dynamic wheelhouse poster or the electronic booklet.

The module is used to perform calculations to predict the path for specific actual or planned commands. In this way the module can be applied to plan and optimise the return manoeuvre and automatically produce the complete situation-dependent manoeuvring plan for a return manoeuvre.

5 SITUATION-DEPENDENT MANOEUVRING PLAN FOR RETURN MANOEUVRE

5.1 Aim and Objective of the Planning Process

The objective of the simulation-based manoeuvre planning and optimisation process is to find a suitable procedure which can be used in a particular situation for the actual status of a real ship.

There are standard files for manoeuvre control settings for simulating specific manoeuvres. By means of the fast time simulations, various results of manoeuvres will be generated. The final goal is to achieve the sequence for an optimised manoeuvre control setting adapted to the actual situation parameter. Presently, the biggest problem is that there are many options possible and the effect of the changes of the parameters used in the models is not very clear; some changes may even have effects which counteract the results of the others. Therefore it is very important to know which parameters which have a clear impact on the manoeuvring characteristic.

An example is given below to indicate the need and the effect of manoeuvring optimisation by means of an Emergency Return Manoeuvre.

5.2 Planning of an Emergency Return Manoeuvre

The example discussed in the following extract is the emergency return manoeuvre using the well known "Scharnow-Turn".

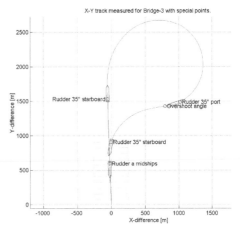

Figure 5. Reference outline for the Scharnow–Turn

As with all other emergency return manoeuvres, the fundamental aim is to return the vessel to the original track by the shortest route and with minimum loss of time. In practice the vessel initially follows the turning circle, and after shifting the rudder by a course change of about 240°, finally turns to counter rudder and amidships. The vessel then swings back to the opposite course at a certain measurable distance from the original track, at a certain distance from the reference manoeuvre.

The first problem is how to get the "Optimal reference manoeuvre" because the heading change of 240° is an average only and can differ among ships from 225° up to 260° or even more, as can the Williamson Turn which can vary from 25° to 80° instead of the standard average value of 60°.

The following figure demonstrates the wide variety of the outcome of the standard course of rudder commands compared for a container vessel, a cruise ship (blue), roro-passenger ferry (brown) and two container feeder vessels (green and red).

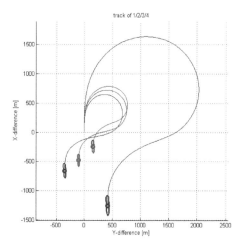

Figure 6. Comparison of the outline of standardised Scharnow–Turn for four different types of ships

Beside this basic variance according to the ship type, there are other more important dependencies that have a substantial impact on the outlined path of a return manoeuvre.

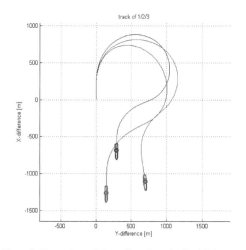

Figure 7. Comparison of the outline of standardised Scharnow–Turn for a 7.500 TEU container ship in ballast condition for three different wind conditions (no wind- blue and wind Bft 6 from north (red) and north-west (green) respectively)

Further samples are given in Fig. 7, which demonstrates the dependency of the final outcome of the return manoeuvre on the loading condition as well as on wind force and wind direction. Of course, the outline would change again if the ship is fully laden or if shallow water effects occur.

Finally, for reasons of completeness, it should be mentioned that there are dependencies on the initial ship speed and on the available water depth. It is clearly to be seen that adaptation of the manoeuvre plan has to be performed for each single varied situation parameter. On the other hand, the simulation software module is able to provide the corresponding data accordingly.

The next step after having simulated the standard manoeuvre procedure for the prevailing environmental circumstances is then to determine the best manoeuvre sequence.

Using the simulation software module there are two principal ways available in order to determine the optimal sequence for the situation dependent manoeuvre plan:

– The first option is to simulate series of manoeuvres using standard „SCHARNOW-Turn" (or WILLIAMSON-Turn) manoeuvring commands in automated simulation series. This method can be seen in Fig. 8 below, where several heading changes were used as parameters to vary the final result of the distance between the initial track and return track.

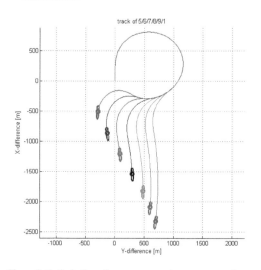

Figure 8. Optimisation of a emergency return manoeuvre by series with different heading changes from 240° up to 300° (with increasing steps of 10°) for counter rudder

The results presented in Figure 8 are for the 7.500 TEU container ship in ballast conditions and taking into account northerly winds of Bft 6.

– The second option is to start with a standard „SCHARNOW-Turn" manoeuvring command series for automated simulation, combined with an optimisation procedure.

An optimising algorithm is applied to find a suitable heading change for counter rudder as parameter to achieve smallest distance (limit=10m) between initial track and return track on opposite heading (limit=2°). The Optimal track is indicated by yellow colour in Fig. 9. The main parameters of the optimised manoeuvre procedure are given in the table format.

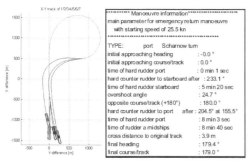

Figure 9. Emergency return manoeuvre optimisation procedure (left) and display of manoeuvring details for optimised manoeuvre (right)

An optimising algorithm is used to find the suitable heading change for counter rudder as parameter to achieve smallest distance (limit=10m) between initial track and return track on opposite heading (limit=2°). The optimal track is indicated by yellow colour in Fig. 9. The main parameters of the optimised manoeuvre procedure are given in the table format.

6 SUMMARY, CONCLUSIONS AND OUTLOOK

Investigations into the overall situation regarding onboard manoeuvring assistance and into the integration of new maritime technologies onboard ships are performed. The ongoing investigations have shown that there is potential to increase operational safety in shipping.

Taking into account the availability of new technologies and new equipment, situation dependent manoeuvring information should be provided to the navigators on the bridge rather than continuing to provide them with static manoeuvring data which often are incomplete and inconvenient in use.

For these purposes, the introduction of a dynamic wheelhouse poster and an electronic manoeuvring booklet is suggested, to provide ship's command with up-to-date information about the manoeuvring characteristics of their ship, adapted to the prevailing environmental conditions.

A concept is developed and exemplarily applied in order to support the accomplishment of manoeuvring tasks in case of a person overboard accident.

The fundamental element of this concept is based on innovative fast-time simulation technologies. It is applied for the purpose of providing situation-dependent manoeuvring data by taking into account actual environmental conditions and actual ship status information. The use is also demonstrated exemplarily for the generation of optimised situation dependent manoeuvring plan for an emergency return manoeuvre.

Future investigations, i.a., will deal with enhancement and validation of suitable visualization of the fast-time simulation results to support decision-making in an ECDIS environment. Therefore, human factor related investigations dealing with a user-centered design of the human-machine interface have to be performed.

Additionally, investigations into the application of the concept on other situations will be carried out.

ACKNOWLEDGEMENTS

The investigations and the preliminary results presented in this paper were partly carried out and achieved within Swedish-German RTD project ADOPTMAN. They belong to the MARTEC program supported by the European Commission. The project is funded and supervised by the Swedish Governmental Agency for Innovation Systems (VINNOVA) and the German Research Centre Jülich (PTJ). Some parts of the work were carried out in the research project "Identification of multivariable parameter models for ship motion and control" (MULTIMAR) funded by the German Federal Ministry of Economics and Technology and the Ministry of Education and Research of Mecklenburg-Pomerania.

REFERENCES

Baldauf, M., Benedict, K., Gluch, M., Kirchhoff, M., Schröder, J.-U. (2010). Enhanced simulation technologies to support maritime operational risk management onboard ships. *Journal of Marine Technology and Environment*, 3(1): pp. 25-38

Benedict, K. et al: (2006) *Combining Fast-Time Simulation and Automatic Assessment for Tuning of Simulator Ship Models*. MARSIM - International Conference on Marine Simulation and Ship Manoeuvrability, Terschelling, Netherlands, June 25th – 30th Proceedings, M-Paper 19 p. 1-9

Benedict, K., Baldauf, M., Fischer, S., Gluch, M., Kirchhoff, M. (2009) *Manoeuvring Prediction Display for Effective Ship Operation On-Board Ships and for Training in Ship Handling Simulators*. IAMU 10th Assembly & Conference 2009 St. Petersburg / Russia at AMSMA, 19-21 September, 2009

Benedict, K.; Hilgert, H. 1986a. *Returning a ship in the case of person overboard accidents*. Part 1 (in German) HANSA, Hamburg, 1986.

Benedict, K.; Hilgert, H. 1986b. *Optimising man-overboard manoeuvres.* 15th Conference of Bulgarian Ship Hydrodynamic Centre, Varna, Proceedings Vol. 1, 1986

IMO 2007. *Revised performance standards for integrated navigation systems (INS).* MSC.252(83). London: International Maritime Organization.

IMO 2009. *Development of Model Procedure for Executing shipboard Emergency Measures.* STW 41/12/3. London: International Maritime Organization.

Lloyd, M. 2007a. *Man overboard. 1:Peparation. Seaways* (2007) April, pp. 22-24

Lloyd, M. 2007b. *Man overboard. 2:Executing the plan. Seaways* (2007) May, pp. 27-28

Jutrovic, I. 2010. *Man overboard. Seaways* (2010) February, pp.3-4

Weintrit, A. 2003. Voyage recording in ECDIS. Shipborne simplified version of Voyage Data Recorders (VDRs) for existing cargo ships based on potential of ECDIS, 11th IAIN World Congress *'Smart Navigation – Systems and Services',* Berlin, 21-24 October.

DISCLAIMER

The views expressed in this paper are the views of the authors and do not necessarily represent the views of IMO, WMU, or the national Authorities.

4. Concept for an Onboard Integrated PNT Unit

R. Ziebold, Z. Dai, T. Noack & E. Engler
Institute of Communications and Navigation, German Aerospace Centre (DLR), Neustrelitz, Germany

ABSTRACT: A robust electronic position, navigation and timing system (PNT) is considered as one of the core elements for the realization of IMO-s (International Maritime Organization) e-Navigation strategy. Robustness can be interpreted as the capability of an integrated PNT system to provide PNT relevant data with the desired accuracy, integrity, continuity and availability under consideration of changing application conditions and requirements. Generally an integrated PNT system is a composite of service components – like GNSS, Augmentation Systems and terrestrial Navigation Systems – and an on-board integrated PNT Unit, which uses the available navigation and augmentation signals in combination with additional data of sensors aboard to provide accurate and robust PNT information of the ship. In this paper a concept of such an on-board integrated PNT Unit will be presented, which is designed to fulfill the specific user requirements for civil waterway applications. At first, the user requirements for an integrated PNT Unit will be overviewed. After that, existing integrity monitoring approaches will be analyzed. Finally, a first integration scheme for an integrated PNT Unit will be presented with a special focus on the internal integrity monitoring concept.

1 INTRODUCTION

The maritime integrated PNT System (Figure 1) is the sum of satellite-based, ashore and aboard components. The integrated use of these components enables the accurate and reliable provision of position, navigation and timing data to all maritime applications.

Position fixing systems are identified as one strategic key element of e-navigation [1]. Existing and future Global Navigation Satellite Systems (GNSS) like GPS, GLONASS and GALILEO are fundamental infrastructures for global positioning. Additionally, terrestrial services are used or considered as candidates to improve the positioning performance (augmentation services: e.g. IALA Beacon DGNSS, RTK) or to ensure the backup functionality (backup services: e.g. e-LORAN, R-Mode) respectively to GNSS. Due to their interoperability and compatibility these systems can be used alternatively or complementary for positioning, navigation and timing.

The International Association of Marine Aids to Navigation and Lighthouse Authorities (IALA) has introduced the term "integrated PNT device" to describe the on-board part of maritime, integrated PNT system. In [2] the integrated PNT device is described as "a device using any available IMO recognized radio navigation systems simultaneously to provide the best electronic position fix for the ship". Following this definition, the outlined objective of the PNT device is focused on the provision of position information to different applications. Several performance standards for shipborne GNSS and DGNSS receivers were developed and approved by IMO in the last decade: GPS [3], GLONASS [4], DGPS and DGLONASS [5], combined GPS/GLONASS [6], and GALILEO [7]. A logical consequence of this standardization process could be the preparation of a new performance standard for a multi-system radio navigation receiver as core element of the on-board part of the PNT System (Figure 1). A more generally admitted approach can be achieved by the introduction of the PNT Unit.

The on-board PNT Unit aims at the provision of position, navigation and timing data in accordance with specified performance requirements, which change during berth to berth navigation. The core of the on-board PNT Unit is a value-added processing system using available radio navigation systems and services in combination with on-board sensors for accurate and reliable PNT-data provision. The on-board PNT Unit is on the one hand part of the integrated PNT system and on the other hand part of the on-board Integrated Navigation System (INS).

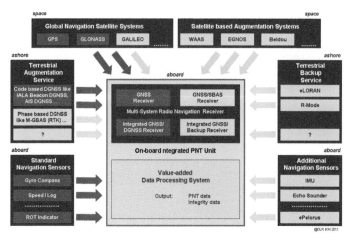

Figure 1. Integrated PNT System (dark grey: standard, light grey: considered options)

Due to user needs such as "Indication and Improvement of Reliability" and "Alarm Management" [1] identified within the framework of the e-Navigation process, a more general approach for the on-board part of the integrated PNT system shall be aimed. Reasons for this perspective are on the one hand the need for redundancy to improve the robustness of PNT-information and to enable the assessment of accuracy by suitable integrity monitoring functions. On the other hand the type of implementable redundancy (equipment, different measurement methods, over determined systems, alternative applicable techniques) specifies the potential of error detection, identification and mitigation.

2 TECHNICAL REQUIREMENTS AND SENSORS

2.1 Technical requirements

In [8] the aim of GNSS is described as a system to provide worldwide position, velocity and time determination for multimodal use. An operational requirement indicates the tasks of shipborne GNSS devices in the provision of position, time, course and speed over ground. But in Appendix 2 and 3 of the same document minimum maritime user requirements are only given for horizontal position. Otherwise the use of the item Electronic Position Fixing System (EPFS) in [9] implicates that the scope of GNSS is rather the provision of position data than the provision of PNT data.

The analysis of these documents shows the necessity to clarify and define the extent of PNT data, which should be delivered by a maritime PNT Unit. In a preliminary design, the following parameters should be considered:

1   Position: It mainly contains the longitude and latitude for maritime navigations. Because vessels can usually be found close to the sea level, the height information is usually not provided as standard output parameter.

2   Under keel clearance (UKC): Instead of the height information, the UKC is the relevant maritime output parameter. It is defined as the distance between the lowest point of the ship (e.g. the keel) and the ground of the sea.

3   Velocity: The magnitude and direction of a velocity vector can be described by Speed over Ground (SOG) and Course over Ground (COG). Because of their physical principles, speed sensors like e.g. electro-magnetic logs can only measure the speed through water (STW), and therefore STW is also a parameter which a PNT Unit could deliver.

4   Attitude: Generally the orientation of the ship in the horizontal plane is reported. Here one needs to distinguish between the orientation with respect to the true north (true heading) and with respect to the magnetic north (heading). For future applications, the other attitude angles, namely roll and pitch, could also be required.

5   Timing: UTC time needs to be delivered.

After the clarification of the PNT output parameters, further additional requirements on the PNT Unit will be discussed in the following.

In [1], the robustness of all e-navigation systems is requested. In order to fulfill this requirement, a definition of robustness needs to be given. We interpret robustness as the ability of a system to provide the output data according to their specification under changing application conditions and in cases of external disturbances (interferences, jamming, atmospheric influences). The robustness shall therefore be applicable to the realization of the basic functionality (output data with required accuracy) and integrity functionality.

In [1] it is furthermore stated, that requirements for redundancy, particularly in relation to position fixing systems, should be considered. Redundancy in a general meaning can be seen as the provision of an alternative system to support critical system functionalities. Within the Recommendation R-129 on GNSS Vulnerability and Mitigation Measures [10] IALA has given a classification of alternative navigation systems in relation to their aims:

1. A **redundant** system provides the same functionality as the primary system, allowing a seamless transition with no change in procedures.
2. A **backup** system ensures continuation of the navigation application, but not necessarily with the full functionality of the primary system and may necessitate some change in procedures by the user.
3. A **contingency** system allows safe completion of a manoeuvre, but may not be adequate for long-term use.

For the introduction of additional sensors to the integrated PNT Unit, this classification scheme needs to be considered.

2.2 *Standard sensors for the provision of PNT data*

According to the carriage requirement demanded by the IMO Safety of Life at Sea (SOLAS) convention [11], following sensors (see Table 1) should be used in maritime applications. An overview of the sensors and a discussion of the related standards can be found in [12]. Therefore here only a table with the list of standard sensors with typical output data and realizations are given.

TABLE 1. STANDARD PNT SENSORS

Sensor	Output	Typical realization
Speed log [13]	Speed through water (STW)	Electromagnetic logs
	Speed over ground (SOG)	Doppler logs
Compass [14] [15]	True heading	Gyrocompass
	Magnetic heading	Magnetic compass
Electronic Position Fixing System (EPFS)	Position Speed over Ground (SOG) Course over ground (COG) Time	GPS/GLONASS DGPS/DGLONASS receiver and antenna
Transmitting Heading Device [16] (THD)	Heading	GNSS multi- antenna system
Echo sounder [17]	UKC	Sonar
Rate of Turn Indicator [18] (ROTI)	ROT	Mechanical Gyro

3 ACCURACY ASSESSMENT AND INTEGRITY MONITORING

3.1 *Integrity Definition*

One of the key tasks of an integrated PNT Unit is the additional provision of integrity messages to the user. Within the specification of requirements for future GNSS [8] IMO has given the following definition of in integrity: The ability to provide users with warnings within a specified time when the system should not be used for navigation. Within that definition the alert limit specifies the applicable threshold differing between an unusable or unusable system. The time to alarm (TTA) describes the acceptable time duration between occurrence of an intolerable error and the provision of the related integrity message. The remaining integrity risk (IR) specifies the tolerable probability, that the assessment of accuracy by integrity monitoring is erroneous. This definition implies an integrity monitoring process wherein the accuracy of the given navigation parameter needs to be estimated in real time and compared against a given threshold (alert limit). In other words, an error estimation of all navigation parameters needs to be performed onboard a vessel.

Integrity monitoring as it is defined for Integrated Navigation Systems (INS) consists of the following three sequential steps [19]:

– Plausibility check
 The plausibility check tests whether the sensor raw data or derived navigational result falls into predefined value range. The plausibility check is normally carried out at receiver side in order to test the usability of the sensor data.
– Validity check
 The validity is tested by comparing the sensor data or derived navigational results with formal and logical criteria as well as by checking the correctness of the data format. The validity check is normally carried out at the sensor site in order to ensure the proper operation of the sensor.
– Compatibility check
 Once a specific parameter can be provided by more than one sensor, different sensor data can be compared to test the compatibility. A significant discrepancy between different sensor data implies the failure of at least one of these sensors. The upper bound for deviation should be defined either a priori or in real-time according to the previous measurements. The compatibility test should be carried out before sending the sensor data to integration algorithm.

The primary aim of plausibility, validity and compatibility checks is the detection of errors. The assessment of accuracy requires the implementation of suitable integrity monitoring functions.

3.2 Existing Integrity Monitoring approaches

In this section existing integrity monitoring approaches defined for maritime navigation will be described and the seen gaps will be discussed.

3.2.1 Integrity monitoring for GNSS

3.2.1.1 Receiver Autonomous Integrity Monitoring (RAIM)

RAIM is a technique, which can be applied in GNSS receivers to assess the integrity of navigation signals [8], if more than 5 GNSS signals are tracked [8]. RAIM has two-fold tasks, first is to check the occurrence of a failure, and second is to identify erroneous satellites. RAIM can be done by using only the incoming measurement of current epoch or employing also measurements of previous epochs. The former approach needs the redundant measurement, and the performance is dependent on the number of satellites in view. In [20], several classic "snapshot" algorithms are introduced.

For a maritime GNSS receiver a RAIM is requested [21], however, neither the algorithm is specified nor it is defined how the user should react on the three possible outputs: safe, caution, and unsafe. Referring to a "snapshot" RAIM, the a priori knowledge of the observation errors is needed by all existing algorithms. A proper determination of this term for maritime navigation is an open task.

3.2.1.2 IALA Beacon DGNSS

GNSS integrity-monitoring services are usually part of augmentation services which also provide DGNSS corrections. The reason for this shared activity is the similarity of the infrastructure required for DGNSS and integrity monitoring.

The IALA Beacon Differential GNSS service is a standardized technique for maritime use [22]. The implemented integrity monitoring assesses primarily the integrity of the service itself, but can also include parts of GNSS assessment. At reference station different thresholds are applied to the observed HDOP or determined range and range rate corrections to estimate the usability of service or single GNSS signals. Additionally the IALA Beacon DGNSS is equipped with one or more integrity monitoring stations operating as virtual user at known location. Only in case that the observed DGNSS position error is below the allowed error threshold, the IALA Beacon DGNSS distributes a flag bit to indicate the usability of the service. The health status of a satellite is provided indirectly by embedding the "do not use" flag in the transmitted PRC and RRC for an unhealthy satellite.

Spatial decorrelation of provided range and range rate corrections, differences in satellite visibility between station and user sites, and user specific reception conditions (multipath, interferences) are the main causes for discrepancies between the estimated and real DGNSS positioning performance.

3.2.1.3 Maritime GBAS (RTK)

The preliminary integrity monitoring of the phase based Maritime GBAS [23] follows the IALA Beacon DGNSS concept. Hence an integrity monitoring station (IM) is installed in the service area (ca. 10-20km) of the reference station (RS). At both stations integrity monitoring procedures are processed in three steps. In the first step the RS and the IM evaluate the quality of the received GNSS signals on the basis of quality parameters like phase and code noise. In a second step a GNSS RAIM based position determination is realized. The results of both steps are compared with pre-specified thresholds to assess the usability of single GNSS signals as well as the usability of the service of RS and IM station.

Afterwards the signals without abnormalities from the RS and IM are used to calculate the IM position by carrier phase based differential techniques (RTK). The position error is derived by comparison of the computed with the exactly known position of the IM. If the accuracy requirements of IMO for port vessel operation are fulfilled, the service can be considered as usable.

Finally the M-GBAS logically combines all gathered integrity information at RS and IM site gained in the previous steps to generate the RTCM 3 messages for the provision of augmentation data and related integrity data (message 4083).

3.2.1.4 Integrity monitoring for INS

An INS offers integrated and augmented functions to support system tasks like collision avoidance, route planning and route monitoring. Currently an INS is not a mandatory system, but if an INS is installed onboard a vessel it is accepted monitoring is considered as an intrinsic function of the INS. The currently valid INS standard is based on IMO resolution MSC.86 (70) [24] and is specified within the IEC-61924 standard [19]. A task oriented concept is already introduced in a new resolution MSC.252 (83) [25], but the specification within the related IEC standard is not yet published. Therefore our analysis is based on IEC-61924 standard only. In IEC-61924 standard, plausibility, validity and compatibility check approaches are introduced. IEC-61924 standard suggests the compatibility check for the following navigational parameters.

1 Position: comparison with a second EPFS; using RAIM GNSS function; Dead Reckoning (DR) using the ship's heading and speed measuring device

2 Heading: comparison with a second heading sensor and a course over ground sensor

3 SOG: comparison with a second SOG sensor, with speed through water sensor and with SOG from the EPFS (GNSS)

4 Time: comparison with a second time sensor and with the internal INS clock

5 UKC: comparison with a second depth sensor and with data derived from ships position and electronic navigational charts (ENC)

3.3 *Demand on specification and development*

The Position, as the most important PNT information, currently is measured by only two separate receiver/antenna GNSS devices. Integrity monitoring is restricted to a comparison of the positions determined by these two receivers. Due to the same measurement technique, system and propagation errors of both GNSS devices underlie the similar error in the measurement and position domain.

An estimation of the actual position error is currently not performed onboard a vessel. Indirectly it is assumed, that the horizontal position error is < 100 m, when using GNSS standard positioning service, and <10 m, when using IALA Beacon DGNSS service.

Although positioning accuracy requirement on GNSS are specified with respect to different operational areas [26][8], the integrity monitoring is currently performed by using fixed or user selectable thresholds. In order to use these areas for future integrity monitoring applications, these areas and the intersection from one area to another need to be clearly specified in an appropriate way (e.g. in ENC charts).

Furthermore integrity specifications for the other PNT parameter (e.g. SOG, COG, time) besides position have to be specified. The necessity of an operational area dependent accuracy and integrity specification for these parameters needs to be discussed

Analyzing the existing integrity monitoring approaches with respect to the identified user needs [1] a demand on the development of an enhanced integrity monitoring for all relevant PNT data within an integrated PNT Unit can be deduced. Such a PNT Unit should use sensor and data fusion methods to ensure the provision of PNT output data with the desired accuracy and to ensure an overall integrity monitoring for these output data. For both functionalities a higher degree of standardization is desired in order to achieve comparable results for their harmonized application.

4 PROPOSED ARCHITECTURE OF INTEGRITY MONITORING IN PNT UNIT

For a PNT Unit, integrity monitoring can be carried out in three sequential steps. The first step is an individual sensor data test. The second step is the compatibility test of similar data from different sensors. The third step is the fault detection and identification in the integration algorithm. A general integrity monitoring approach is depicted in Figure 2.

4.1 *Integrity monitoring for GNSS*

Actually within an INS, position integrity monitoring is performed by comparing the calculated position with a second EPFS, by using RAIM GNSS function and by using DR technique. Possible improvements in a PNT Unit will be elaborated as follows.

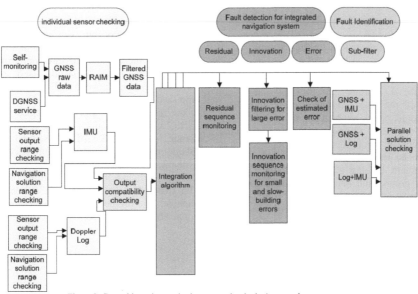

Figure 2. General integrity monitoring approaches in the integrated sensor system

4.2 Compatibility check for redundant GNSS systems

GNSS redundancy can be achieved using a secondary GNSS device, using multiple civilian frequencies and using multiple GNSS constellations.

Once the major GNSS device (antenna or receiver) is out of use, the second GNSS device can fully take the function of the major GNSS device. However, the redundant GNSS device is also affected by the errors related to the radio signal. In this sense, the significance of a redundant GNSS device is reflected during the internal failure of the major GNSS device.

Modern GNSS satellites will facilitate more than one signal. Civilian code data will also be encoded into the carrier signal besides L1 signal at future GNSS satellites. The additional civilian code data will offer the same functionalities like the SPS service. Due to different carrier signals, the other carrier signals might not suffer from the same interference or jamming or propagation effects as L1 signal. Also, the channel failure (loss of lock or cycle-slips, etc.) for L1 signal might not occur simultaneously on the other frequencies. Nevertheless, errors due to space atmosphere and signal propagation will influence all the carrier signals of a satellite. Hardware failure of receiver or antenna might also challenge the reception of all carrier signals.

Two or more full-operational GNSS constellations could serve as redundancy for each other, as they realize same functions in maritime navigation as specified in [26]. However, GLONASS and GALILEO are not yet fully operational. Future GNSS receivers and corresponding antennas allow the reception and processing of multiple GNSS signals, however, a hardware failure can cause the loss of all GNSS signals.

4.3 Compatibility check with backup systems

Systems like e-Loran (enhanced Long Range Navigation) or R-mode (Ranging mode) facilitate the functions for positioning. The e-Loran system can also be used for time determination, so that these systems could serve as backup for GNSS [27]. Compared to GNSS, e-Loran signals are transmitted at lower frequency with higher power and hence it is not easy to be jammed especially not by the same GNSS jammers. It relies on the radio signal propagated over ground and hence does not suffer from the same errors in the propagation path from sky like GNSS. So the future of e-Loran as a terrestrial backup for GNSS with a large coverage area is currently an open question. Also, the fulfillment of the future maritime requirements with respect to the accuracy is an issue.

In [2], the R-Mode is seen as a possible novel variant of positioning technique using terrestrial signals. The idea is to use existing communication channels and append their functionality by sending an additional timing signal. From the time difference between signal transmission and reception, the ship should be able to determine its position. The advantage of this idea would be that at least partially existing infrastructure could be used. Currently this is still only an idea, where the proof of concept needs to be shown.

4.4 Compatibility check with contingency system

The DR is a frequently-used technique to predict the position using SOG and COG information. In maritime navigation, COG information is usually approximated by compass. DR is independent of the radio signals and hence still works during GNSS outage. In a sensor fusion system, DR is already an implicit function and does not need to be separately implemented.

Another contingency system can be constructed by introducing the inertial sensor. Also the inertial sensors are still not standard sensors in maritime navigation, they are drawing more and more interests due to the independency of radio signals, the short-term high accuracy, the high output rate and the decreasing price. An Inertial Measurement Unit (IMU), which is composed of three orthogonal accelerometers and gyroscopes, can offer the navigation parameters like position, velocity, rate of turn and attitude. For this reason, introducing an IMU allows the integrity monitoring for relevant sensors. A drawback is that the IMU cannot work alone for long-term use and hence needs to be integrated with other sensors.

4.5 RAIM

Classic RAIM algorithms can also be applied to the maritime navigation. The only problem is the determination of the a priori measurement error for pseudoranges. This relies on standardized specification for different operation areas. If this is not available, empirical values have to be used. As an added value of the integrated system, an enhanced RAIM aided by the antenna dynamics can be implemented. The classic "snapshot" RAIM is based on the received pseudorange data of current epoch. Once the antenna dynamics can be determined, the antenna position can be predicted from the position of last epoch. This makes the pseudoranges of current epoch predictable. The predicted pseudoranges serve as additional observation to enhance the integrity monitoring.

4.6 Integrity monitoring for other independent navigation sensors

Except for GNSS, the error estimation for navigation sensors is difficult as the sensor raw data is not directly processed. However, the sensor fusion algorithm makes it possible. Once the error estimation of one sensor data is available, the fusion algorithm allows the error estimation for other relevant sensors. For example, if the error of GNSS- based SOG can be properly assessed, the error analysis of speed log is also possible.

4.7 Integrity monitoring in integration algorithm

The plausibility tests, validity tests and compatibility tests are suitable for detecting gross sensor failure but not sensitive for slight error, time-variant errors and drifts. The Kalman filter-based algorithm could offer high sensitivity of detecting these errors. Integrity monitoring based on Kalman filter can be categorized into the following approaches [27].

4.7.1 Kalman filter estimates (bias check)

In a Kalman filter, the errors of navigation parameters can be estimated. If an estimated error is significantly larger than the error level specified by the manufacturer, it is likely to be a failure in the sensor.

4.7.2 Innovation-based approaches

The innovations indicate the consistency of the actual measurements and the measurements predicated by state estimates. Innovation filtering may be used to detect large discrepancies immediately, whereas innovation sequence monitoring enables smaller discrepancies to be detected over time.

4.7.3 Residual-based approaches

The above-mentioned innovation filtering and sequence monitoring can also be expanded to residuals. Residuals have a smaller covariance than innovation, making them more sensitive for error detection [27]. The only shortcoming is that the processing of residuals is not an essential part of a Kalman filter routine and needs extra computing time.

4.7.4 Parallel solution of multiple sub-filters

Parallel-solutions integrity monitoring maintains a number of parallel navigation solutions or sub-filters, each excluding data from one sensor or radio navigation signal. Each additional navigation solution is compared with the main filter using a consistency test. A significant inconsistency indicates a fault in the sensor or signal omitted from main filter. The system output is then switched to the solution omitting the faulty sensor or signal. The main drawback lies in the increased computational burden and hence this technique is preferably used for failure identification rather than failure detection

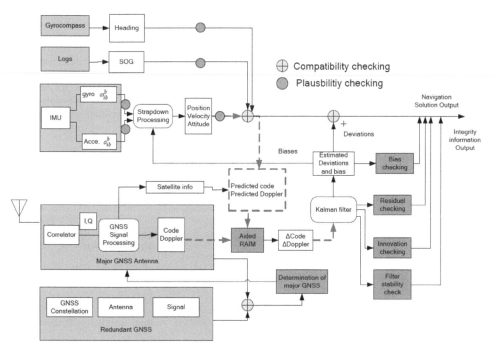

Figure 3. Integrity monitoring in the architecture of a PNT unit

4.8 Initial design of integrity monitoring in PNT Unit

The major integration strategy of a PNT Unit is the integration of GNSS and IMU. An initial design of sensor integration in a PNT Unit is depicted in Figure 3. The plausibility check and compatibility check are marked at the corresponding positions in the data flow.

The plausibility check is carried out for gyrocompass, speed log and IMU output. Taking the dynamic properties of a specific ship into account, the threshold value can be derived for plausibility check.

The compatibility check is first applied to GNSS in order to determine the major GNSS antenna (also the constellation and signal, if not only one constellation or one signal is to be used). The data of major GNSS antenna will be integrated with other sensors. Both GPS and IMU offer velocity parameters and hence can be compared with speed log. IMU also outputs attitude (heading) and hence can be compared with gyrocompass.

The Kalman filter enables the bias check, innovation check and residual check. The prerequisite is a stable operation of the filter mechanism. As a stochastic system, the filter performance is based on the modeling of the observation and dynamic model, the reasonable *a priori* knowledge of the observation and dynamic errors, and most important, the real application scenario. It is not a trivial task to adapt the filter to all potential unexpected situations, and hence it is necessary to test the stability of the filter.

The stability test can be done either using the internal parameters of the filter, or by checking the compatibility of filter results with the results from other stand-alone sensors.

4.9 Integrity output from a PNT Unit

According to the previous analysis, the following integrity parameters will be supported by a PNT Unit.

TABLE 2. OUTPUT FROM A PNT UNIT

PNT Output	Plausibility	Validity	Compatibility	Estimated Error
Position	x	x	Redundant GNSS Other radio-navigation system	x
SOG	x	x	GNSS Doppler, Log, IMU,	x
COG	x	x	GNSS, IMU	
Attitude	x	x	Compass multi-antenna system	x
Rate of Turn	x	x	IMU heading variation with time	x
Time	x	x	Other time sensors internal clock of an INS	

5 SUMMARY

This paper focuses on a maritime integrated PNT Unit as the on-board part of maritime PNT system. The aim of the PNT Unit is the robust provision of position, navigation and timing information in accordance with the performance requirements of the actual operational region. IALA has introduced the term PNT device [2] as "a device using any available IMO recognized radio navigation systems simultaneously to provide the best electronic position fix for the ship". The PNT Unit, proposed in this paper, follows a more general approach in combining available radio navigation systems and their augmentation services with other on-board PNT sensors. The core of the PNT Unit is a processing system, which combines all available PNT sensors. The PNT Unit is on the one hand part of the integrated PNT System and on the other hand part of the on-board INS.

Focusing on integrity for PNT information we have analyzed the state-of-the-art integrity monitoring approaches with respect to the identified user needs. Based on this, a demand on the development of an enhanced integrity monitoring for all relevant PNT data within an integrated PNT Unit can be deduced. Such a PNT Unit should use sensor and data fusion methods to provide PNT output data and improve integrity monitoring for these output data. For both functionalities a high degree of standardization is desired in order to achieve comparable and reliable results for their harmonized application.

Subsequently we have introduced a preliminary integrity monitoring concept for a PNT Unit which also includes additional sensors in order to deliver redundancy, backup or contingency functionality.

Finally it should be stated that this paper can only be seen as a starting point towards the realization of an integrated onboard PNT Unit for maritime applications. In a next step, after consolidation of the architecture, we plan to develop a demonstrator system of a PNT Unit.

REFERENCES

[1] IMO, NAV 54/25 Annex 12 Draft strategy for the development and implementation of e-Navigation, 2008.
[2] IALA, World Wide Radio Navigation Plan, Edition 1, 2009.
[3] IMO, Resolution MSC.112(73): Adoption of the Revised Performance Standards for Shipborne GPS Receiver Equipment, 2000.
[4] IMO, Resolution MSC.113(73): Adoption of the Revised Performance Standards for Shipborne GLONASS Receiver Equipment, 2000.
[5] IMO, Resolution MSC.114(73): Adoption of the Revised Performance Standards for Shipborne DGPS and DGLONASS Maritime Radio Beacon Receiver Equipment, 2000.

[6] IMO, Resolution MSC.115(73): Adoption of the Revised Performance Standards for Shipborne GPS/GLONASS Combined Receiver Equipment, 2000.

[7] IMO, Resolution MSC.233(82): Adoption of the Performance Standards for Shipborne Galileo Receiver Equipment, 2006.

[8] IMO, Resolution A.915(22): Revised Maritime Policy and Requirements for A Future Global Navigation Satellite System (GNSS), 2001.

[9] IMO, NAV 56/WP.5 Development of an e-Navigation Strategy Implementation Plan, 2010.

[10] IALA, Recommendation R-129: On GNSS Vulnerability and Mitigation Measures Edition 2, 2008.

[11] IMO, SOLAS Chapter V: Safety of Navigation, 2002.

[12] R. Ziebold, Z. Dai, T. Noack, und E. Engler, "Concept for an Integrated PNT-Unit for Maritime Applications," Proceedings of the 5th ESA Workshop on Satellite Navigation Technologies. Navitec 2010, 8.-10. Dec. 2010, Noordwijk, The Netherlands.

[13] IMO, Resolution MSC.96(72): Adoption of Amendments to Performance Standards for Devices to Measure and Indicate Speed and Distance, 2000.

[14] IMO, Resolution A.424(XI): Performance Standards for Gyro Compasses, 1979.

[15] IMO, Resolution A.382(X): Magnetic Compasses Carriage and Performance Standards, 1977.

[16] IMO, Resolution MSC.116(73): Performance Standards for Marine Transmitting Heading Devices (THDs), 2000.

[17] IMO, Resolution A.224(VII): Performance Standards for Echo- Sounding Equipment.

[18] IMO, Resolution A.526(13): Performance Standards for Rate-Of-Turn Indicators, 1983.

[19] IEC, IEC 61924 Maritime navigation and radiocommunication equipment and systems – Integrated navigation systems – Operational and performance requirements, methods of testing and required test results, 2006.

[20] B.W. Parkinson und J.J. Spiker, Global Positioning System: Theory and Applications: Volume II, American Institute of Aeronautics & Astronautics, 1996.

[21] IEC, Maritime Navigation and Radiocommunication Equipment and Systems – Global Navigation Satellite Systems (GNSS) – Part 1: Global Positioning System (GPS) Receiver Equipment, Performance Standards, Methods of Testing and Required Test Results.

[22] IALA, Recommendation R-121: The Performance and Monitoring Of DGNSS Services in the Frequency Band 283.5 – 325 KHz, 2004.

[23] D. Minkwitz und S. Schlueter, "Integrity Assessment of a Maritime Carrier Phase Based GNSS Augmentation System," in Proceedings of ION GNSS, Portland, USA: 2010.

[24] IMO, MSC.86(70): Adoption of new and amended performance standards for navigational equipment, 1998.

[25] IMO, Resolution MSC.252(83): Adoption of the Revised Performance Standards for Integrated Navigation Systems (INS), 2007.

[26] IMO, Resolution A.953(23): World-Wide Radio Navigation System.

[27] P.D. Groves, Principles of GNSS, Inertial, and Multi-Sensor Integrated Navigation Systems (GNSS Technology and Applications), Artech House Publishers, 2007.

ECDIS

5. A Harmonized ENC Database as a Foundation of Electronic Navigation

M. Bergmann

Jeppesen, a Boeing Company, Frankfurt Am Main Area, Germany

ABSTRACT: The current discussion on the future of electronic navigation is focusing on the expansion of ECDIS by integrating additional data streams, like AIS or real-time tide information. While this are important aspects, which will be touched on in the paper, it is important to note that the ENC data layer is the necessary data layer to enable advanced data display.

The current focus on ENC production has a limited view on coverage within scale bands and on datum code. While this allows concentrated efforts to produce the necessary country coverage to meet the timeframe of the IMO ECDIS mandate, future mariners will require a more solid basis of chart data for their electronic navigation needs. The IMO e-Navigation discussion and the discussions in related the IALA e-Navigation Committee are starting to join efforts with the new IHO S-100 and S-101 standards as well as the developing S-10X series of standards. The Hydrographic community needs to take that in account when preparing for the future of electronic navigation to increase safety of navigation in a rapidly changing environment with larger ships and more traffic in areas with specific navigational challenges.

The paper will focus on the following topics:
- The ENC data layer as the basis of electronic navigation,
- Closing gaps and overlaps in ENC coverage by adjusting cell boundaries,
- Moving from a cell based data structure to a seamless database structure,
- Integration of Scale-Independent and Scale-Dependent objects,
- Adaption of harmonized and flexible data models – an S-10X outlook,
- Enabling integration of advanced data streams – an e-NAV outlook.

1 THE LANDSCAPE OF ELECTRONIC NAVIGATION

In different organizations, involved in safety of life at sea, aspects of navigating a vessel from port to port are discussed. The importance of navigation is highlighted by the International Maritime Organization (IMO) in the "Safety Of Life at Sea" (SOLAS) regulation, especially in chapter V. The discussion on safety of navigation is key to the IMO Maritime Safety Committee (MSC). Because MSC realized this important aspect of SOLAS, it created the "Safety of Navigation" Sub-Committee (NAV) focusing on exactly this.

The discussion in NAV centered for years around the "Electronic Chart Display Information System" (ECDIS). In 2010 MSC, on request of NAV, has approved a new regulation, which outlines the mandates of ECDIS on defined SOLAS-Class ships in a phased-in approach starting 2012 and reaching all desired vessels by 2018.

While IMO focuses on the regulatory aspects to ensure and improve safety of navigation at sea, the International Hydrographic Office (IHO) since its establishment in 1921 focuses on the cartographic aspects of safety of navigation as well as associated activities.

On its websites the IHO defines its goals as follows:
"The object of the Organization is to bring about:
- The coordination of the activities of national hydrographic offices,
- The greatest possible uniformity in nautical charts and documents,
- The adoption of reliable and efficient methods of carrying out and exploiting hydrographic surveys,
- The development of the sciences in the field of Hydrography and the techniques employed in descriptive oceanography (www.iho-ohi.net).

❘❘ TIMETABLE FOR ECDIS CARRIAGE REQUIREMENTS			
Ship type	Size	New ship	Existing ship
Passenger ships	≥500 gross tons	1 July 2012	No later than 1st survey after 1 July 2014
Tankers	≥3,000 gross tons	1 July 2012	No later than 1st survey after 1 July 2015
Dry cargo ships	≥50,000 gross tons	1 July 2013	No later than 1st survey after 1 July 2016
	≥20,000 gross tons (new ships) 20-50,000 gross tons (existing ships)	1 July 2013	No later than 1st survey after 1 July 2017
	≥10,000 gross tons (new ships) 10-20,000 gross tons (existing ships)	1 July 2013	No later than 1st survey after 1 July 2018
	3-10,000 gross tons	1 July 2014	No retrofit requirements to existing ships <10,000 gross tons

Key components of these objectives are the standards for the "Electronic Nautical Charts" (ENCs). Especially the current S-57 standard defines those ENCs, which are, per ECDIS performance standard, the only vector data sets, which allow operating an ECDIS in its so called ECDIS mode, and as such to perform primary navigation with electronic systems.

IMO has looked at the Hydrographic Offices (HOs) around the world, represented by IHO, to provide adequate ENC coverage before adapting the ECDIS mandate. The IHO has confirmed to IMO-MSC and IMO-NAV that by 2012 adequate ENC coverage will be available.

All of the above highlights that both IMO as well as IHO see ENCs in ECDIS as the foundation and primary data set for electronic navigation.

2 THE "E-NAVIGATION" CONCEPT

Following its initiative on ECDIS, IMO has launched another initiative: "e-Navigation". The "e-Navigation Correspondence Group" identifies details on "enhanced Navigation", which is intended to integrate shore based and ship based systems and data streams to increase situational awareness in any phase of sailing and as such increase Navigational safety even further. The discussion circles around concepts like integrated AIS (Automated Identification System), different system overlays on board, automated information exchange between shore and ship, like sharing Vessel Traffic System (VTS) data and so on.

The IMO e-Navigation development and the discussions in the related e-Navigation Committee of the International Association of Marine Aids to Navigation and Lighthouse Authorities (IALA) are starting to join efforts. Within the last few months various working groups have taken great efforts to align

the development in IHO, IALA and IMO and to harmonize the view of the future of electronic and enhanced navigation. In those meetings the organizations, the industry and individual expert contributors could reach agreement that the new IHO S-100 and S-101 standards and concepts as well as the developing S-10 X series of standards should be used to build the bases for the new marine data models. This includes the usage of the IHO "GI-Registry", the "Geographical Information Registry", which supports the harmonization of GIS-Data Models across the maritime industry and its initiatives. This development provides the necessary harmonized platform for integrated systems. The hydrographic community starts to take the initiative preparing for the future of electronic navigation to increase safety of navigation in a rapidly changing environment with larger ships and more traffic in areas with specific navigational challenges.

The concept of e-Navigation is evolving the understanding that future navigation will need constant innovation, and as such will need to change how performance standards are handled. It is widely understood that the current ECDIS performance standard is restricting innovation. Its update and certification concept is not geared up to meet the needs of e-Navigation.

The currently being developed new concept defines a framework in which a growing number of data streams are integrated and harmonized to allow the creation of the necessary information for increased Situational Awareness in an environment of growing complexity.

The on board and on shore systems to be developed within this framework will have to create a compelling need for their usage by increasing safety and security of navigation (compelling need for coastal administration) and improved efficiency of voyage (compelling need for ship owners and operators).

The dominant argument unifying all stakeholders to move towards common structures and towards the IHO originated model is the fact that the implementation of S-100 and its related standards is well underway and will materialize shortly. As a consequence ENCs will follow this data structure, so are associated data streams, like "Inland ENCs" or "Marine Information Overlays". As all stakeholders agree that the ENC layer will build the foundation of any kind of advanced navigational systems, it was a natural development to try to align other data stream with this foundation. But looking from the other side this development again verifies that ENCs, or better Hydrographic Vector Chart Data Layers, are the necessary ingredients for any navigational display now and in the foreseeable future.

3 MIGRATION FROM A "CHART CENTRIC" TO A "SITUATIONAL CENTRIC" CONCEPT

The traditional hydrographic work to create the necessary tools for mariners to navigate safely is utilizing classic cartographic concepts.

An early chart from the year 1603 illustrates that cartographic art work is used to allow the knowledgeable navigator gaining sufficient information for a safe passage.

In the "paper" or "analog" world, this has developed over hundreds of years as the best practice to transport the necessary information. The current paper charts of HOs around the world are in most cases beautiful artwork, well developed to help navigate ships.

Even in the electronic world, this concept started to materialize with the usage of "Raster Charts" in displays, with ARCS (Admiralty Raster Chart Service) as a prominent example. The stakeholders then realized that the full potential of electronic navigation cannot be explored with this raster charts. Vector cartography showed new opportunities of increased situational awareness by utilizing capabilities of data links, full zooming capabilities without "fading" and a growing number of other advantages. While the first vector charts had been developed unregulated by key stakeholders in the industry, the now already often mentioned ENCs have made their way on bridges of SOLAS class ships as the only official electronic navigational chart.

While this development was the necessary next step in electronic navigation, the current ENCs, based on the S-57 standard, are not reaching far enough. These ENCs are "cell based", which means they are still looking at a certain "chart", a defined rectangle on the globe. As a consequence, what is today offered is not a real "Hydrographic Database", but rather a collection of associated charts in a central data repository. The HOs try to harmonize the cells to create a kind of "seamless" appearance in the ECDIS display. But as the view of each cell in the creation is still often a "chart by chart" view, this harmonization is not always successful. An addition complication in harmonizing such a chart centric view on ENCs is the fact that the current system is in general focusing on a "scale band" concept. Here the hydrographic data is composed by cartographers in a certain cell to be for optimal use on a certain zoom level. As this is on a cell by cell and scale band by scale band level, harmonization is not only necessary between cells of the same scale, but also across scale bands. Because of the high level of complexity this harmonization is mostly omitted. Data conflicts and as such display conflicts are the results when moving from cell to cell on a "moving map" display as a ship travels, but it also creates conflicting information as a navigator is zooming in or zooming out and with that moves the focus of the ECDIS from one independently developed scale band to another scale band.

As we can see the chart centric ENC production process, while it generates a great improvement, is almost impossible for the process to generate an increased situational awareness with as little confusion as possible to the mariner. Even within the responsibility of one single HO. As SOLAS ships have the tendency to cross borders, the ECDIS systems are dealing with data sets from different countries. This increases the complexity and in consequence issues like overlapping data, data gaps or mismatching of adjacent cells.

Where "Regional ENC Coordination Centers" (RENCs) are used, those RENCs are also trying to help harmonizing the ENCs, but of limited success, given the complexity of the task.

The current focus on ENC production has a limited view on coverage within scale bands and on datum code. While this allows concentrated efforts to produce the necessary country coverage to meet the timeframe of the IMO ECDIS mandate, future mariners will require a more solid basis of chart data for their electronic navigation needs. The above discussed data issues needs to be addressed and resolved in order to gain confidence of mariners in electronic navigation.

The issues I have highlighted are well known and endless discussions around the globe took place and are conducted right now trying to find solutions. S-100, fully developed, will help mitigate some of those risks. S-100 will move the hydrographic data collection towards a GIS (Geographical Information System) oriented data concept, away from cell based thinking. The future concept will move more and more data sets from scale dependant data storage towards the scale independent data concept. In this concept a natural object, like a buoy or a shallow area, will be stored in its real-world location and dimensions. The database will contain the object only once and the rendering engine will compose the display, based on detail rendering and deconfliction rules, rather than using an artwork-like cartographic view.

In a full database centric hydrographic data collection, cell boundaries may still exist, but they are created to break up the data sets in manageable pieces, but will no longer be developed as individually composed data sets. The hydrographic data will be manager as a complete set, rather than individual puzzle pieces, which are stitched together like a patchwork quilt to create an individual coverage. This underlying data layer will be rendered based on situational needs, i.e. zooming level as desired by the mariner, and as such will result in a situational centric display.

4 CONCLUSION

The current development of usage of electronic cartography in the maritime world has taken a step towards situational awareness and as such has matured away from simple chart display. This development will intensify and as such will require in future a change in how electronic maritime cartography is developed, composed and stored. The future will focus on data streams to support data integration and situational centric rendering.

In addition the financial and organizational pressure on HOs will require optimization of workflows and better utilization of capacities.

To support these changes a harmonized database of hydrographic data, a harmonized ENC database is paramount.

REFERENCES

IMO, 2009, SOLAS (Consolidated Edition)
Jeppesen, 2010, ECDIS – What you need to know
IHO, 2000, IHO Transfer Standard for Digital Hydrographic Data, Edition 3.1
IHO, 2010, IHO Universal Hydrographic Data Model, Ed 1.0.0
IHO, 2011, www.iho-ohi.net
IHO, 2011, Operational Procedures for the Organization and Management of the S-100 Geospatial Information Registry, Ed 1.0.0

6. Navigation Safety Assessment in the Restricted Area with the Use of ECDIS

Z. Pietrzykowski & M. Wielgosz
Maritime University of Szczecin, Szczecin, Poland

ABSTRACT: This paper presents an analysis of vessel safety parameters used in the ECDIS system while navigating in restricted areas. Apart from defining their priorities, a group of parameters indispensable for the safe navigation in restricted waters is identified.
The function of ship domain is proposed on the basis of safety parameters defined in the ECDIS system. This function may be utilized in the navigation decision support system that uses ECDIS data and included as a new function in the ECDIS system.

1 INTRODUCTION

The year 2012 will be the first year of mandatory installation of the Electronic Chart Display and Information Systems (ECDIS) onboard ships. The first installation requirements refer to newly built vessels, depending on their type and size. Then in the years 2016-2018 the regulation will come in force for ships in service. The navigator of the vessel under the mandatory requirements of the SOLAS Convention will receive an essential tool changing the rules of navigation, watchkeeping and assessment of ship's safety. This tool has alarm and indicator functions that provide important aid to the navigator. Its proper use requires a valid user-defined parameters of safe navigation and activation of selected alarms. A large number of applicable safety features causes difficulties in its use in navigation along a desired route. Therefore, it is advisable to specify a group of basic safety parameters as a minimum for the planning and safe monitoring of sea passage.

The analysis of alarms and related safety parameters was carried out on with ECDIS NaviSailor 3000i device by Transas Ltd. (Transas, 2004a, Transas, 2004b, Grzeszak et al. 2009)

It should be noted that not all alarms and safety parameters dealt with are mandatory according to the performance standards for ECDIS systems (IMO Resolution A.817/19 1995, IMO Resolution A.232/82 2006, Weintrit 2009). Some of them are introduced by manufacturers of such systems as part of enhancing their functionality.

2 ALARMS IN ECDIS

2.1 *Types of alarms*

There is a large number and variety of alarms, so they can be classified according to various criteria. The authors propose the division of alarms according to these criteria: 1) priority of the alarm, 2) possibility of activation and deactivation, 3) source, 4) the scope of the alarm, and 5) basic / others.

The first criterion divides the alarms into: a) alarms, b) indications . This division results from the provisions of IMO Resolution A.817(19), A.232(82) and IMO "Code on Alarms and Indicators" (IMO-867E). The state of the system requiring attention and action is signaled by an alarm in the form of acoustic or acoustic and optical signal. The state of the system requiring attention mainly of the user, without having to take immediate action, is indicated in the form of an optical indication only.

Among alarms implemented in ECDIS system the following are distinguished (criterion 2): a) the alarms that cannot be deactivated (e.g. Safety Contour, Depth Safety), b) alarms that can be deactivated (e.g, Sounder Depth, Anchor Watch), c) alarms that can be deactivated, but the user responsible for the safety protects them with a password. These alerts are activated in different ways, and some require implementation of the safety parameter in advance.

The division of alarms due to the source (criterion 3) includes alarms by: a) hardware, sensors) b) system. The former signal states of disability or reduced functionality of devices, including sensors of the system and the system as a whole, such as networking. The other group contains alarms associated with the implementation of functions relating to navigational situation. They signal a significant event for the safety of navigation.

The fourth division includes the proposed classification of alarms according to the criterion of scope of activities. This term is understood as the functions of alarms associated with the types of threats. These can be distinguished: a) antigrounding alarms b) alarms associated with the route of the ship – "Route alarms" c) Target / Radar alarms d) Area type alarms or "Area alarms" , e) other alarms; f) AIS alarms g) alarms and indications related to the scale and type of the chart.

Experiments conducted at the Maritime University of Szczecin during model ECDIS courses, resulting in issuing the ECDIS operator's certificate, show that course participants do not use many system capabilities, and also have problems with the interpretation of alarms and indications. This is mainly due to the different specifics of the work on the ENC as compared to working with paper or raster charts. Lack of understanding by the operator of the principles of interpretation of ENC content by the ECDIS system results in significantly reduced utilization of the system, and even the use inconsistent with the idea of the system. This involves the use of ECDIS system on the principles applied to classical paper charts, where interpretation of the contents of the chart lies belongs to the user only. It is connected with the fact in that during the process of navigator training primarily paper charts are still used. A better use of ECDIS systems requires, therefore, wider use of ENC in the training of navigators.

Table 1. Alarms classification criteria

	Criterion			
Alarm type	Activation/ deactivation	Source	Scope of activities	Priority
1	2	3	4	5
a) alarm	deactiv. impossible	equipment, sources	anti-grounding	basic
b) indication	deactiv. possible	system	route	other
c)	deactiv. possible (password-protected)		target	
d)			area alarms	
e)			others	
f)			AIS alarms	
g)			chart alarms	

Significant help in the correct use of the system may be an additional division and allocation of alarms and indications (criterion 5): a) basic alarms and indications, necessary for safe voyage monitoring b) other, complementary to the previous one.

These criteria and the classifications of alarms are summarized in Table 1.

2.2 Basic and other alarms

Taking into account the above, an analysis of alarms and indications was performed according to the criterion of ECDIS scope of activities (criterion 4), with the proposal and explanation for the division into basic and other alarms (criterion 5).

The alarms in question are presented in Table 2., categorized by the groups of alarms identified in the ECDIS NaviSailor 3000i system.

Table 2. Groups of alarms according to the presented criteria (see Table 1).

Group of alarms/ indications	Location (ECDIS NaviSailor 3000i)	Criterion				
		1	2	3	4	5
antigrounding alarms	monitoring/ nav. alarms	a, b*)	a, b, c*)	b	a	a
	system	a, b*)	a, b, c*)	b	a	a
route alarms	monitoring/ route mon.	a	a, b*)	a, b*)	a, b*)	a, b*)
	system	a	a, b*)	a, b*)	a, b*)	a, b*)
target/ alarms	targets/ ARPA	a	a, b*)	a, b*)	a, c*)	a, b*)
	system	a	a, b*)	a, b*)	a, c*)	a, b*)
areas/ basic areas	monitoring nav. alarms	a	a	b	d	a
areas/ add. areas	monitoring nav. alarms	a	a	b	d	b
other alarms	monitoring/ nav. alarms	a	a	b	e	a, b
	system	a	a	b	e	a, b
	config.	a	a	b	e	a, b
AIS alarms	alarms/ AIS alarms	a	a	b	d	a
chart alarms	system	a, b*	b	b	g	a, b*)
	charts	a, b*	b	b	g	a, b*)
	monitoring.	a, b*	b	b	g	a, b*)

*) due to the diversity of alarms in the group, it was necessary to assign some of them to more than one group according to the criterion.

The group of "Antigrounding Alarms" contains: a) Nav. danger, b) Safety contour changed, c) Anchor watch, c) Safety contour, e) Safety depth, f) Ag monitoring off, g) Safety scale changed.

The group of "Route Alarms" contains: a) Off chart, b) End of route, c) Out of XTE, d) Behind schedule, e) Ahead of schedule, f) WP approach, g)

Course difference, h) Prim / Sec diverged, i) Chart datum unknown, j) Prim. not WGS 84, k) Sec. not WGS 84, l) Track control stopped, m) Backup navigation, n) Low speed, o) Dangerous drift, p) Course change.

The group of "Target / Radar Alarms" contains: a) –CPA / TCPA, c) Lost target, c) Guard zone target, d) Disk full save reset, e) Disk full adjust save, f) Head marker failure, g) Bearing failure, h) Trigger failure, i) AIS message.

The group of alarms "Area Alarms" contains 28 "Basic Areas" alarms and 14 "Additional Areas " alarms.

The group of "Other Alarms" contains: a) Timer went off, b) End of watch, c) Time zone changed, d) No official chart, e) Add info warning, f) Add info chart full.

The group of "AIS Alarms" contains: a) Tx malfunctioning, b) Antenna VSWR exceeds limit, c) Rx channel1malfunctioning, d) Rx channel 2 malfunctioning, e) Rx channel 70 malfunctioning, f) general failure, g) MKD connection lost, h) External EPFS lost, i) No sensor position in use, j) No valid SOG information, k) No valid COG information, l) Heading lost / invalid, m) No valid ROT information.

The group of "Chart Alarms" contains: a) Dangerous scale, b) Not recommended scale, c) Layers lost, d) Look up for better chart, e) Larger scale chart available, f) ENC data available, g) Chart priority / HCRF mode, h) Safety scale / check on larger scale than, i) No official chart (also included in "Other Alarms".

Basic alarms are considered as alarms which are important for the safety of sea passage. Among others, they include: commonly used collision warning, sounder depth alarm, lost target, cross track error - XTE. The newly introduced alarms for ECDIS systems which work on the basis of vector charts were considered as important. These include safety contour, safety depth, area alarm, navigational danger.

Selecting the basic alarms may facilitate their activation, and editing the safety parameters associated with them.

3 NAVIGATION SAFETY PARAMETERS

The effectiveness of alarms depends on the proper definition of safety parameters associated with them. These efforts should include the nature and circumstances of the area of navigation. A necessary condition is also their selective activation (except for system alarms), taking into account the type of area and navigational situation. Another problem, not analyzed in this article, is the selection of alarms to be activated reflecting the experience and knowledge of a specific sea area by the navigator.

The analysis highlights the safety parameters associated with the movement of the vessel on the surface and in the third dimension - depth and underwater hazards.

3.1 *Navigation safety parameters associated with the movement of the vessel on the water surface*

These parameters apply to both fixed and mobile objects that threaten the safety of navigation, also including parameters related to the navigation accuracy and maintaining the vessel's position and route.

CPA, TCPA. The basic parameters of this group are the Closest Point of Approach (CPA) and Time to Closest Point of Approach (TCPA), edited by the navigator and activating collision warning alarm. They cover both AIS targets (if the presentation is switched on) and the radar / ARPA objects. When the ARPA is connected to the ECDIS as its sensor, the limit values of these parameters can be edited independently in the ECDIS and the ARPA. In this case, the ARPA is treated as a system sensor for the ECDIS, which means that the alarm from system ARPA must be repeated in the ECDIS system.

Guard Ring (Rings), Guard Zone (Zones). Important safety parameters are the radius of the area of automatic acquisition or parameters that define the zone or zones of automatic acquisition. In the latter case they may be the values of angle sectors with inserted distance from the unit. These parameters, similarly to the parameters of CPA and TCPA, can be edited independently in the ARPA and the ECDIS.

Area Vector. This is a vector representing time from the intersection of area type objects. Time setting, selected by the navigator, is represented as a vector calculated on the basis of a calculated COG and SOG. It may be displayed together with the Safety Vector. The navigator has a choice of area objects. The selection of area objects should be done depending on the crew experience and knowledge of the sea area.

RMS Circle (Root Mean Square Error Circle - at 95% confidence level). The parameter is calculated automatically by the system, but the novelty of increasing its relevance and practical use in ECDIS is introduced in a graphical presentation of the circle and the expected trajectory. This allows to verify the setting of parameter XTE.

Limit of Cross Track Error - XTE. This parameter is independently defined by the navigator to the left and right side of the route and may be different at each leg of the route. Its importance lies in the fact that it determines the width of the ship trajectory checked by the system (by using the "Check" function, checking whether the planned route exceeds or not the safety parameters set by user. Importantly, the automatic system " Track Control " (mandatory in ECDIS) will try to keep the ship within this limit.

Divergence in the primary and secondary positioning system. This parameter allows the navigator to monitor the difference in position of the vessel obtained from two positioning systems (generates optional alarm "Primary / Secondary Diverged"). In addition to activation, the user sets a limit of the distance at which positions from the main and secondary positioning systems can diverge.

Parameters related to the monitoring of the ship's planned route. These are course difference, WP approach, out of schedule, off chart, last WP passed. Incorrect setting of these parameters may cause a navigational accident.

Parameters describing the activation of functions related to charts to be used: priority of loaded charts (Chart Priority), and automatic chart loading "Chart Autoload". Improper use of these features can result in a lack of alarms specific to the ECDIS system (working with the charts other than ENC S-57 standard charts).

3.2 Navigation safety parameters associated with underwater hazards

Safety Contour. This parameter, defined by the navigator, is one of the most important safety parameters of modern navigation. The parameter possible for use in ECDIS only when the system uses vector charts. It generates an alarm of intersecting the safety contour. If the chart does not have a selected safety contour in its database, the system automatically sets the next higher (safer) contour.

Safety Depth. Parameter defined by the navigator. It extends the ability to detect underwater hazards found at depths greater than specified by safety contour.

Time from the intersection of safety contour - Safety Vector. The parameter is defined by the navigator. It allows designation and presentation of the "Safety Vector" on the basis of calculated COG and SOG.

Sounder Depth. Parameter defined in the echo sounder as a sensor of ECDIS system. This means that the alarm will be repeated in the ECDIS system. This allows the verification of depth, read from the ENC.

Navigational danger ring radius. It defines a circle of safety "Navigational Danger Ring". This enables the detection of underwater hazards on the basis of ENC. It also allows detection of: a) Navtex objects that have the attribute "Danger" added, b) objects inserted by the user as "User chart object" or "Manual correction object" which was given the attribute "Danger" and / or inserted depth less than the "Safety Depth ".

3.3 Other safety parameters

Safety scale. Parameter defined by the user. It determines the chart scale for checking safety contour and safety depth. It means that the system will monitor underwater hazards on charts with a scale larger than that determined (Fig. 1).

Differential mode lost. The time for which signal is lost from the DGPS reference station. The excess value of this parameter generates an alarm. This is important because of the decline in accuracy of fixing the position from DGPS to GPS.

Display category imaging. This parameter defines the scope of the presented on-screen navigation information: Base, Standard, Custom, All.

Shallow Contour and Deep Contour (Fig. 1). The parameters are defined after the function "Four Shades" is activated to present additional areas of shallow water (Shallow Contour) and deep water (Deep Contour). The function modifies the displayed chart by creating four depth areas with different colors.

Fig. 1 Safety parameters window (safety contour, safety depth and safety scale)

3.4 The basic safety parameters

The analysis of alarms and safety parameters, their location in the ECDIS system and the consequent difficulty of access to them makes it advisable to introduce the function "Basic Safety Parameters Settings". This function would allow the operator, in one tab or window, to define and monitor the basic safety parameters, and activate alarms necessary to ensure safe voyage realization by the ship equipped with ECDIS. It should include viewing and editing, and the activation state of alarm associated with them. These are:

1 safety contour,
2 safety depth,
3 safety scale,
4 chart display category,
5 „chart priority",
6 CPA/TCPA,
7 cross track error – XTE,
8 course difference,
9 WP approach,
10 safety vector (advance in the intersection safety contour),
11 area vector (advance in the intersection of area objects),
12 Navigational Danger Ring, its radius,

13 chart display and ship's motion (North Up, Head Up, Course Up, Relative Motion, True Motion)
14 difference in the position of the vessel from primary and secondary positioning system (Primary / Secondary diverged),
15 presentation of AIS targets (on / off),
16 presentation of ARPA objects (on / off),
17 special areas detecting defined (yes / no),
18 presentation of the COG (course over ground) and COW (course over water) vectors (on / off),
19 off chart (on / off).

Before the start of a voyage or when the system is restarted, ECDIS should automatically require the operator to define or confirm the values of these safety parameters with the possibility of automatic switch to the window where the operator activates and edits that alarm.

4 SHIP DOMAIN

4.1 *Ship domain as a safety criterion*

Safe operation of the ship requires constant analysis and evaluation of the situation. On this basis navigator undertakes decisions concerning navigation. The analysis and assessment of the situation are carried out in accordance with the criteria adopted by the navigator. A commonly used criterion in collision avoidance systems is the closest point of approach. However, in the case of navigation in restricted waters, particularly in narrow fairways and channels, it is difficult to apply in most cases. This is due to the lack of free choice of route and the need for compliance with safety rules, taking into account local conditions (restriction of one of the three dimensions defining the distance of the ship from other objects).

An alternative to the mentioned criterion of the navigational safety is the criterion of the ship domain. Application of the criterion of ship domain enables quick identification and assessment of the navigational situation and thus developing the decision support in ship's maneuver. It should be noted that this criterion is also possible to use in the open sea areas. For example, this criterion was implemented in the prototype of navigational decision support system for seagoing vessels developed at the Maritime University of Szczecin (Pietrzykowski et al.2009).

The concept of ship domain was introduced in the 1970s (Fuji & Tanaka 1971, Goodwin 1975). It is assumed that the domain is an area (domain two-dimensional) or space (three-dimensional domain) around the vessel which should be kept clear of other objects.

Assuming a certain level of discretization of relative bearings (eg $\Delta \angle K = 1°$), the domain boundary of the vessel B_{DS} is described by a curve passing through the n points p_{Di} ($i = 1,2,..., n$), located on the relative bearings $\angle K_i$ at the distances d_{DSKi} from centre of the vessel (eg, centre of waterline):

$$B_{DS} = \{p_{D1}, p_{D2}, ..., p_{Dn}\} \qquad (1)$$

Ship domain boundary D_S at different bearings is then described as follows:

$$D_S(\angle K_i) \le d_{DSKi} \qquad i = 1, 2,..., n \qquad (2)$$

The basic problem is to define the domain boundary, dividing the area around the ship into sub-areas: dangerous and safe. It is a difficult task because the shape and size of the domain are affected by many factors. These include: size and maneuverability of the vessel, parameters of the area where the ship maneuvers, hydro-meteorological conditions, vessel speed and the speed of other vessels, the intensity of vessel traffic in the area, the accuracy of position fixing, training level, knowledge and experience of navigators. Also significant is the adopted method of determining the ship domain boundary.

The issue of determining the domain was presented in many publications, including (Fuji & Tanaka 1971, Goodwin 1975, Coldwell 1983, Zhao et al. 1993, Smierzchalski & Weintrit 1999, Pietrzykowski 2008, Pietrzykowski & Uriasz 2009, Wang et al. 2009). There are two-and three-dimensional domains proposed in the literature. The former describe the area around the ship. Domains of two-dimensional shapes include circle, rectangle, ellipse, polygon, complex plane figures. In the case of three-dimensional domains - they describe also vertical space included between ship and sea bottom and the air draft of the ship. Their shape often corresponds to sphere, ellipsoid, cylinder, truncated cone.

Among the methods of determining the ship domain one can distinguish three groups: statistical methods, analytical methods and artificial intelligence methods. It is characteristic for all these methods that they make use of navigators' knowledge, both procedural and declarative.

Application of statistical methods requires the registration of relevant data. In addition to difficulties in collecting them, the problem that arises is to separate various factors that influence the shape and size of the domain.

Analytical methods are based on the analytical description of the domain space. These methods ensure precise description of the ship domain. The main difficulty is to take into account and balance all relevant factors affecting the shape and size of the domain.

Methods of artificial intelligence (AI) were developed to acquire and use the knowledge of expert navigators using the tools of artificial intelligence. They include and use, *inter alia*, fuzzy logic, artificial neural networks and evolutionary algorithms.

4.2 Possibility to define a domain based on the safety parameters in the ECDIS

Safety parameters available in the ECDIS system do not define directly a ship's domain. These authors analyzed the possibility of identifying the two-and three-dimensional domain using the parameters analyzed in the article. The problem was brought down to the determination of the length, width and shape for two-dimensional domain, and in the case of three-dimensional domain additionally the depth and shape of the geometric solid.

If we use the CPA parameter value to determine the length of the domain D_L then the length of the domain takes the value

$$D_L = 2CPA \qquad (3)$$

This results from the fact that this parameter defines a safe distance at which other vessels pass, is widely used, and its interpretation is unambiguous.

Due to the difficulty in determining the safety parameter indicating the width of the domain designation was proposed based on the analytical relationship between the length and width of the domain. This relationship can be derived on the basis of ship domain analytical descriptions proposed, *inter alia*, in (Coldwell 1983, Zhao et al. 1993, Smierzchalski & Weintrit 1999, Pietrzykowski 2008, Pietrzykowski & Uriasz 2009, Wang et al. 2009):

$$D_W = f(D_L) \qquad (4)$$

The simplest figure describing the domain of the ship on the basis of the parameters (D_L, D_W) is a rectangle. Taking into account the results of statistical research on the shape of the domain, the domain was proposed in the shape of an ellipse inscribed in a rectangle with sides (D_L, D_W).

The parameter BCR- bow crossing range can be an alternative to the CPA safety parameter, used for describing the length and, consequently, the width of the domain.

ECDIS system gives definitely a lot more opportunities for determining the domain of the third dimension - depth D_D (three-dimensional domain). The parameter defining the third dimension of the domain can be safety contour or properly set safety depth. It seems to be necessary to use safety depth, which results from a broader range of hazards analyzed by the system for that parameter.

Then a three-dimensional domain is described as a solid with two bases in parallel planes. The upper base is an ellipse (two-dimensional domain). The bottom base is a circle defined by the radius D_R of navigational danger ring. The circle origin is an orthographic projection of the ellipse origin. The side surface of a geometrical solid is a section of the plane connecting the two bases with the smallest surface area (Fig. 2).

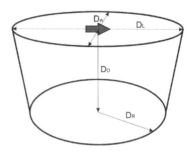

Fig. 2 Three-dimensional domain

When the domain function is implemented in the ECDIS, the domain parameters (D_L, D_W, D_D, D_R) will be generated automatically as a default with the possibility of correction by the navigator (like other safety parameters).

5 CONCLUSIONS

Based on analysis of alarms and indications of ECDIS system and safety parameters defined by the navigator, the group of basic parameters necessary for the safe sea passage was proposed particularly for use in restricted areas. These parameters will be available after activating the "Basic Safety Parameters Settings" in an additional window. This allows the navigator to set alarms, activate them and define the safety parameters necessary to ensure safe sea passage of the ship equipped with ECDIS. When the solutions herein proposed are implemented by manufacturers and positively verified by navigators in practice, it will be recommendable to consider options for revising the performance standards for ECDIS systems.

Due to the limited capacity of the CPA parameter to be used in the safety assessment when navigating in restricted areas, these authors considered the possibility of defining the ship domain as a safety criterion in the ECDIS system. The definition of two- and three-dimensional ship domain based on the safety parameters defined in the ECDIS system is proposed.

REFERENCES

Coldwell T., 1983, Marine Traffic Behaviour in Restricted Waters. Journal of Navigation 36: 431-444.

Fujii Y., Tanaka K., 1971, Traffic capacity, Journal of Navigation 24: 543-552.

Goodwin, E.M., 1975. A statistical study of ship domains. Journal of Navigation 28: 328-344.

Grzeszak J., Bąk A., Dzikowski R., Grodzicki P., Pleskacz K., Wielgosz M. 2009. Przewodnik operatora systemu ECDIS Navi-Sailor 3000 ECDIS-i. Szczecin: Wydawnictwo Naukowe Akademii Morskiej.

IMO Resolution A.232/82, 2006. Revised Performance Standards for Electronic Chart Display and Information Systems (ECDIS), International Maritime Organization, London.

IMO Resolution A.817/19, 1995. Performance Standards for Electronic Chart Display and Information Systems (ECDIS), International Maritime Organization, London.

Pietrzykowski, Z., 2008. Ship's Fuzzy Domain – a Criterion for Navigational Safety in Narrow Fairways, Journal of Navigation 62: 499-514.

Pietrzykowski, Z., Uriasz J., 2009. The ship domain – a criterion of navigational safety assessment in an open sea area. Journal of Navigation 61: 93-108.

Pietrzykowski Z., Magaj, J., Chomski J., 2009. A navigational decision support system for sea-going ships, Measurement Automation and Monitoring PAK 10: 860-863.

Transas, 2004a. Navi-Sailor 3000 ECDIS-I. Version 4.00.01. Operating Principles, Transas Ltd..

Transas, 2004b. Navi-Sailor 3000, ECDIS-I, Version 4.00.01. User Manual, Transas Ltd.,.

Smierzchalski, R, Weintrit, A., 1999, Domains of navigational objects as an aid to route planing in collision situation at sea. In: Proc. of 3rd Navigational Symposium, Gdynia, I, 265-279, (in Polish).

Wang N., Meng X., Xu Q., Wang Z., 2009. A Unified Analytical Framework for Ship Domains. Journal of Navigation, 62: 643-655.

Weintrit A., 2009, The Electronic Chart Display and Information System (ECDIS), An Operational Handbook, CRC Press Inc. Taylor and Francis Group, p. 1101.

Zhao J, Wu Z., Wang F., 1993. Comments on ship domains. Journal of Navigation 46: 422-436.

7. Increasing Maritime Safety: Integration of the Digital Selective Calling VHF Marine Radiocommunication System and ECDIS

M.V. Miyusov, V.M. Koshevoy & A.V. Shishkin
Odessa National Maritime Academy, Ukraine

ABSTRACT: A general project for simplification of VHF DSC radiocommunication for navigators and for increasing DSC efficiency by integration of the DSC VHF radio equipment and the AIS – ECDIS vessel's system is presented. Proposed measures are to contribute to further developments in maritime radiocommunication systems, technology and progress of e-navigation strategy.

1 INTRODUCTION

Technical progress of modern digital navigating and communication technologies has provided new technical systems appeared and maintained in practical navigation such as: Global Maritime Distress and Safety System (GMDSS), Automatic Identification System (AIS), Global Navigational Satellite System (GNSS), Electronic Chart Display and Information System (ECDIS), etc. At the same time keeping and increasing the level of safety at sea in the conditions of vessel's saturation by navigating and radioelectronic devices demands the effective operation of the specified systems under common fatigue of the officer of the watch (OOW).

Not always the practical navigational tasks can be satisfactory provided within the existing shipborne equipment. In particular there is misalignment between decision time and reaction time. It is necessary to provide real time operational procedures.

In Odessa National Maritime Academy at the Electrical Engineering and Radio Electronics Faculty, (Maritime Radiocommunications chair) in the general direction of safety navigation enhancement and development of electronic navigation (e-navigation) since the end of 90th years scientific-practical works are carried out in the following directions:

1 Improving of the GMDSS Narrow-Band Direct Printing interface for telex communication;
2 Radar detection of high-speed vessels with small surface of reflection against interferences from a water surface;
3 Automatic identification of VHF radiotelephone transmissions in real time;
4 Increasing of DSC efficiency by means of integration with AIS and ECDIS ship systems.

In the present report some results of researches in the fourth direction are presented. The obtained results have been formulated and submitted by Ukraine in the form of the technical proposal for discussion of 14th Session of IMO Subcommittee on Radiocommunications, Search and Rescue (COMSAR), that have been passed in London in February, 2010 [1]. The given proposal was unanimously supported by the participants of the Session.

Digital Selective Calling (DSC) is one of the basic features of GMDSS radiocommunication subsystems. In accordance with International Telecommunication Union Recommendation [2] all radiotelephone transmissions of any priority (distress, urgency, safety and routine) must be preceded by the proper digital selective call. Nevertheless, the procedures of radiocommunication with the use of DSC are often neglected either in cases of distress or with other priorities. In particular, VHF channel 16 is often used incorrectly for any calls attributed with routine priority as it was foreseen in the old system procedure instead of using DSC on channel 70.

The reasons for such neglect were analysed in numerous documents of COMSAR, for example [3]. The main reason for the DSC communication procedure neglect is described in this document as the navigators' nonconformity with the DSC procedures based on:

– difficulty of manual DSC forming procedure;
– different DSC interface provided by manufacturers;
– the overload of the DSC call forming menu by secondary, non-important functions; and
– limited space for all DSC format visualization, etc.

Navigators as usual neglect DSC process and directly pick up the telephone on channel 16. Such elimination of DSC doesn't accelerate reliable and correct VHF communication because a lot of time may be needed for called/calling vessel's verbal specification among other vessels.

An important objective of the VHF radiocommunication improvement is the development and implementation of such technical improvements which could give an ability to:

1 simplify the process of providing DSC radiocommunication to, practically, the equivalent actions in terms of time and number of operations, as to the usual radio-telephone receiver picking up procedure; and

2 select the called/calling vessel among others which are displayed on the electronic navigational chart. This element is specifically important for the urgent reaction of the watch officer on the called vessel under difficult navigation conditions, i.e. to provide the automatic identification of the called/calling vessel in the live navigation situation.

The completion of these two tasks can be achieved within the frames of the currently used vessel equipment through the integration of the VHF DSC controller and the navigation equipment composed of the Automatic Identification System (AIS) and the Electronic Chart Display and Information System (ECDIS). In this case all of the basic functions of integrated systems are preserved.

Currently AIS is an obligatory equipment to be carried on board all vessels. ECDIS has been mandated recently, nevertheless even now it is widely used as a supplement to the traditional paper charts. AIS provides for the exchange of information which includes an identification number (MMSI). ECDIS, while operating with AIS, enables it to indicate vessels on a navigation display within the AIS operating area, i.e. approximately 30 nautical miles, and to reflect them on electronic navigational charts.

The combination of the VHF DSC equipment with the AIS - ECDIS system allows:

1 to eliminate, practically, the procedure of DSC forming while substituting manual operations by a computer mouse "click" on the ECDIS display; and

2 to provide the authentication of a calling vessel on the electronic chart and thus to make the process of attachment of the called/calling vessel to the navigation situation automatically. A calling vessel can be indicated on the display by a blinking mark which will allow OOW (officer of the watch) of the called vessel to quickly estimate the navigational situation and make an effective decision.

The unique character of the vessel's authentication is provided by the presence of vessel identifier (MMSI) both in DSC and AIS. In other words, it means that the calling vessel is automatically attached to the current navigation situation represented on a navigation display.

During the manual preparation of the call procedure, as it is provided in the existing system, an operator has to make an individual call to a vessel/coast station and for this purpose has to enter a nine-character long digital identifier (MMSI) and working channel number. While doing this, about 20 pressings of the DSC controller keyboard buttons are required. More difficult calls require more key pressing. The proposed method gives the ability to form a call through a mouse click only on the chosen vessel (or the coast station). The series of parameters, for example, the working channel number, can be set by default (or can be chosen manually if necessary). The entering of the MMSI is not required in the proposed method because it can be sent from the AIS - ECDIS system to the DSC controller automatically.

Integration of the DSC - VHF and AIS - ECDIS may be realized through a separate interface block connected to the DSC equipment of any equipment manufacturers. Replacement of the DSC equipment is thus not required. Appropriate simplified scheme is presented in the figure 1. The scheme includes also navigational sensors and devises that in practice are connected to ECDIS Workstation. Connection of DSC VHF equipment and ECDIS is implemented by means of bidirectional interface.

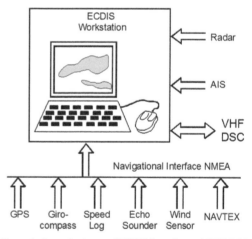

Figure 1. General scheme of ECDIS inputting and VHF DSC connection

Integration of the DSC - VHF and AIS - ECDIS requires no changes to the existing radiocommunication operational procedures. All regular DSC forming and viewing functions are saved. What is essential is that the present manual method of forming/viewing calls will be preserved as a supplementary means to the automatic method of forming/viewing calls in the AIS - ECDIS system.

VHF DSC integration with the AIS-ECDIS will ensure further enhancement of safety of navigation while simplifying the navigator's interface with radiocommunication and navigational equipment and accelerating actions of operator.

The practical problem on the way of such kind of integration is the necessity of ECDIS modernization by means ECDIS - DSC connection and interaction in both directions. Corresponding apparatus means of interface and software tools are developing.

This suggestion is fully compatible with the e-navigation development strategic direction which envisages further development of means of radio-communications and navigation and the implementation of modern digital information technologies in navigation.

Technical implementation of this suggestion is also compatible with the new regulations for the mandatory carriage requirement of ECDIS.

REFERENCES

[1]. Proposal for simplification of VHF DSC radiocommunication and increasing DSC efficiency. Submitted by Ukraine / SUB-COMMITTEE ON RADIOCOMMUNICATIONS AND SEARCH AND RESCUE. 14th session, Agenda item 7. COMSAR 14/7, 27 October 2009
[2]. Recommendation ITU-R M.493 "Digital Selective-Calling System for Use in the Maritime Mobile Service".
[3]. Simplification of DSC equipment and procedures. Submitted by Finland / SUB-COMMITTEE ON RADIOCOM-MUNICATIONS AND SEARCH AND RESCUE COMSAR 8/4/1, 27 November 2003.

8. Enhance Berth to Berth Navigation requires high quality ENC's – The Port ENC – a Proposal for a new Port related ENC Standard

D. Seefeldt

Hamburg Port Authority (HPA), former Head of the Geographic and Hydrographic Department, retired end of 2009
Work Package Leader within the integrated European Research Project EFFORTS (Effective Operation in ports); responsible for the Subproject Port ECDIS

ABSTRACT: The Hamburg Port Authority (HPA) was about 42 month, between May 2006 and October 2009, the work package leader for the Port ECDIS work package within the integrated European research project named EFFORTS (Effective Operation in Ports). The Port ECDIS team was completed by the company's SevenCs (Germany), CARIS BV (The Netherlands) and the ISSUS Martime Logistics / TUHH (Germany). The HPA was responsible for the development of a proposal for a new Port ENC standard which can be used for navigation in ports on board of vessels, in PPU's (Portable Pilot Units), in VTMI-Systems, in a state of the art marine simulator, for port maintenance and other harbor related tasks. Masters and pilots approaching a seaport usually use an Electronic Chart Display and Information System (ECDIS) to obtain the required navigational information they need. Also the Harbor Master needs the same up-to-date information for the admission process and to organize a safe and ease navigation in the port area. The common ECDIS standard supports navigation in the open sea and coastal areas; the Inland ECDIS standard was developed for navigation on inland waterways. The chart requirements for maneuvering big ships in confined waters like narrow fairways (harbor access channels), turning and harbor basins, for port maintenance (dredging), fairway and channel design and construction work, for TUG operation and for traffic management (VTMIS) are not sufficiently covered by the current ECDIS and Inland ECDIS standard with respect to chart scale, accuracy, chart objects, attributes ("object catalogue", in future "feature catalogue") and topicality and call for a special Port ENC. Managing bigger vessels, increasing traffic, less harbor space, berth organization, dredging purposes etc. requires accurate and up-to-date high-resolution geographic and bathymetric data to provide all necessary information. The Port ENC it is not just about producing better electronic charts (the Port ENC or PENC) to be shown in the navigation displays of various applications. Port-ECDIS addresses user groups of other domains as well (maintenance, dredging, planning, simulation, engineering, TUG assistance, VTMIS, voyage or route planning). Often they have the need to look at the data not only as a chart but also in 3D. That means additional data representations are required. The Port ENC must be able to interact with other port related data sources for a more beneficial use and to improve the interoperability of harbor related tasks. The Port ENC could also play a fundamental role in the e-Navigation concept!

1 WHY A PORT ECDIS?!

1.1 *Introduction*

Ports are the hubs of global trade with the need of the highest level of topographic and hydrographic information to fulfil special requests regarding safe and ease of navigation, manoeuvring, turning, docking, berthing et cetera. That takes into account the special requirements Harbor Masters, Pilots, Ship Officers, TUG operators, Transport Execution and Port Maintenance have! This requirement should be fulfilled by the Port Hydrographer! That's a real challenge, because safe and efficient arrival/departure of ships and their cargo is most crucial for ports!

Increase of vessel sizes versus less harbor manoeuvre space, Minimum Under Keel Clearance and special requirements for minimum dredging call for the highest level of accuracy and reliability of digital chart information for navigation in fairways, turning and port basins currently not met by equipment according to SOLAS V Carriage requirements!

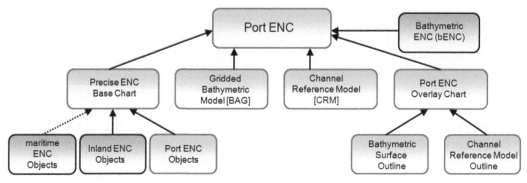

Fig. 1: The Port ENC component

The common IHO ECDIS standard for maritime ENC's supports navigation in the open sea, coastal areas and in seaports (like the Port of Hamburg), the Inland ECDIS standard for Inland ENC's (IENCs) and was developed for navigation on inland waterways and uses the same accuracy and quality definitions like the maritime ECDIS standard, but both without meeting the requirements ports have regarding precise navigational, manoeuvring, berthing, turning, docking, maintenance, up to-dateness, scale and accuracy aspects!

Port ENC requirements go far beyond the current maritime ECDIS and Inland ECDIS standards regarding up-to-dateness, quality, accuracy, large scale charts, chart features/objects and attributes and reliability of hydrographic data (Bathymetry) and geographic data (Topography). For Port operations, there are special requirements for vertical and horizontal accuracy. That is achieved by using modern sensor technology. The same accuracy must be inherent in the underlying electronic charts, the Port ENC's. This type of source data (e.g., topography and hydrographic data) has to be made available by the Port Authorities using a standardized data format, the proposed Port ENC standard, because they are responsible for this task. So the Port Authorities as a kind of public institution should be an approved Port ENC producer, I think!

At present, there is no standard or extensions considering the special requirements of port operations! That calls for a specific "Port ECDIS", the Port ENC standard. The following figure shows the different components of the Port ENC, including gridded bathymetry and a 3D channel reference model and also ready for using the 7Cs bathymetric ENC.

The Port ENC standard should be an independent but complementary standard to maritime ENC and Inland ENC (see below).

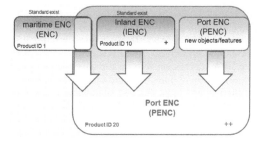

Fig. 2: ENC gradation including the Port ENC

The development of a Port ENC standard focuses on high precision operations in ports. A Port ENC intended to align with the ongoing developments for maritime and Inland ENCs with respect to the new IHO standard S100. And Port ENC data should serve as the missing link between maritime and Inland ENCs, because seaports are often the link between maritime and inland shipping.

Using a Port ENC as the base, it must be possible to overlay other types of information to improve the interoperability of harbor-related tasks, for example navigation and ship – manoeuvring and docking by Pilots using Portable Piloting Units (PPUs) including the Port ENCs. Also the Port Authority can use the Port ENC for dredging and maintenance activities at channels, piers and berths and the same Port ENC can be used as base for traffic management and route planning in the nautical centre (VTMIS).

1.2 *IHO S57 and S44 Standards – comments*

IHO Standards do not provide significant topographic source data for integration in ENC's. No dedicated accuracy requirements are defined that apply for different navigational purposes / categories (e.g., port operations). Within ENC's and Inland ENC's, the IHO S-57 Zone of Confidence (ZOC) assessment is used to describe the quality of bathymetric data, but the Zone of Confidence (ZOC) is not used for topographic data!

ID		S-52
bathymetric	topographic	representation
1	1	
1	2	

Tab.3: S-52 representation for the meta object "Accuracy of ENC data"

Fig. 3: Port ENC encoding guide proposal

The IHO S57 Standard and the latest IHO S44 Minimum Standard for hydrographic surveys should be harmonized in terms of their accuracy data.

As in figure 3 shown, within the Port ENC we combined accuracy arguments for the bathymetric and topographic information and represent these two classes with a Zone A or Zone B symbolisation. The highest accuracy level for a Port ENC (Zone A) has to fulfil IHO S44 Special Order Survey and a horizontal and vertical accuracy for fixed topographical objects relevant for berthing, docking et cetera better than +/- 0,1 m. The second level (Zone B) has also to fulfil S44 Special Order Survey and a topographic accuracy level better than +/- 0,5 m. This is much higher than in the current ECDIS standard!

An example is the official ENC of Hamburg, produced and issued by BSH (Federal Maritime and Hydrographic Agency / Germany). It meets all the relevant ENC related standards and fulfills the requirements for maritime navigation (SOLAS V carriage requirements), but the ENC is too small in scale, does not have any bathymetric detail, does not show up-to-date information and includes poorly defined horizontal accuracy for topographic features such as quay walls, piers, pontoons, et cetera. A comparison of the official maritime ENC and the new developed Port ENC reveals the following: the official maritime ENC is not suitable for special operations within the port area! To be fair, the official BSH - ENC has a different purpose to meet (usage band 5 - harbor), but must be used as official ENC in the Port of Hamburg to fulfil SOLAS V carriage requirements.

2 THE PORT ECDIS WORK PACKAGE – TASK OVERVIEW

Task 1 – Potential user requirements (Meetings, workshops, structured questionnaire)

Task 2 - Port ENC - Technical specification

– accuracy; precision of topography and aids of navigation; special new Port ENC objects (features and attributes); precise 3D depth information using Digital Terrain Models (DTM) technologies; 3D gridded bathymetry, 3D reference DTM (the Channel Reference Model CRM)

Task 3 – Prototype of a Port ENC

– Port ENC dataset of the Port of Hamburg, including precise Port ENC chart data, so named 3D gridded bathymetry (in BAG format), bathymetric ENC's (bENC) and a 3D channel reference model (CRM).

Task 4 – Testing of prototype(s)

– Tests on board of HPA survey vessels; test using a PPU on board of a container vessel, functional tests onboard of a Trailer Suction Hopper Dredger (TSHD) and during docking process of a cruise liner.

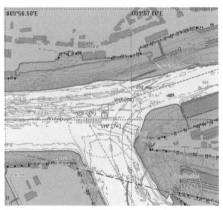

Fig. 4: Port ENC + bENC (Bathymetric ENC)

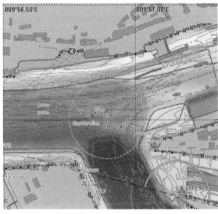

Fig. 5: Port ENC + 3D Gridded Bathymetry

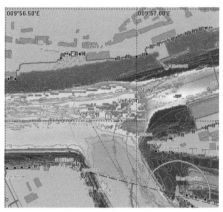

Fig. 6: Port ENC- calculated safety depth

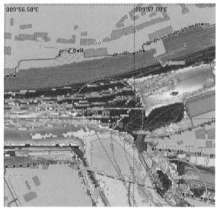

Fig. 7: 3D Gridded bathymetry data versus CRM

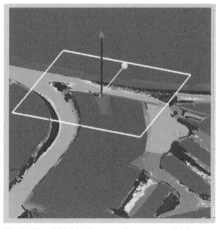

Fig. 8: 3D gridded bathymetry data versus CRM

The figures 4 – 8 giving an overview about the results specified in Task 2 of the Port ECDIS work package.

Task 5 – Defining requirements for follow-up developments and standardization (Port ENC - Roadmap).
- The Port ENC can be used as base information within a PORTIS (Port Information System) which also includes AIS, Radar, VTMIS, Route Planning, dredging information, river and port basin maintenance information, current and velocity, tidal information etc. Follow-up work to enhance the prototype, widen its application and organise standardisation was described.
- Port ENC can also be used in Marine Simulators (ship handling, tug simulator...) et cetera.

3 PORT ECDIS WORK PACKAGE RESULTS - OVERVIEW

- The outcome of the Port ECDIS work package was a proposal and comprehensive concept as basis and input for European / international standardization proved by validation and functional tests in the Port of Hamburg. We produced a paper about the "Definition of present Data Quality in Standards used for ENC data (S57 versus S44 standard) – current situation" and some Port ENC specification documents like a "Port ENC Feature Catalogue", a "Port ENC Encoding guide (representation and symbolisation) and the "Port ENC Product specification". A Port ENC prototype (software and dataset) of the Port of Hamburg including a Port ENC viewer was also developed. We made a lot of very successful tests using the Port ENC prototype (based on basic dataset). All the tests running very successful, delivering very promising results and demonstrating the outstanding quality and accuracy of the developed Port ENC (report)!! At least we wrote a "Port ENC follow-up requirements document".
- The very successful result of the EFFORTS work package 1.3 - Port ECDIS could be only a proposal and comprehensive concept for a new Port ENC standard. It can be currently only a first step.
- The results were distributed to a lot of official bodies like IHO, IMO, IALA, IHMA, IMPA et cetera and should be discussed also within the global Port and Harbor community and with navigation related organizations (currently started). I think main Ports like Singapore, Rotterdam et cetera and also Hamburg should produce their own precise Port ENCs, because they are responsible for safe and ease of navigation in the port area. The Hamburg Port Authority thinks about how to set it up best.

4 CONCLUSIONS

The outcome of the Port ECDIS work package was a proposal and comprehensive concept as basis for European and international standardization proved by validation and functional tests in the Port of Hamburg.

The Geographic and Hydrographic Department of the Hamburg Port Authority produced a very precise and up to date Port ENC test dataset of the Port of Hamburg area which meets the user requirements of the involved nautical participants like Harbor Masters, Pilots and Port maintenance bodies like dredging, TUG operators, marine simulator and others.

The definition and scope of the IMO draft strategy for the development and implementation of E-Navigation focused on marine information on board and ashore by electronic means to enhance berth to berth navigation and related services. So I think, the Port ENC can play an important role and could/must be a core component for E-Navigation

REFERENCES

IHO Publications, Standards and Specifications:
    S52, S57, INT 1, S100 et cetera.
SEEFELDT, D., HOFFMANN, R., ROWAN, E.: Port ECDIS – Enhanced ENC Standard for Port Operations, Hydro International, Vol. 14, Sept./Oct. 2010, p. 19-23, 2010.
Links:
    http://www.hamburg-port-authority.de/presse-und-aktuelles/aktuelle-themen.html
    http://www.efforts-project.org/index.html

9. The New Electronic Chart Product Specification S-101: An Overview

J. Powell
National Oceanic and Atmospheric Administration, Silver Spring, United States

ABSTRACT: The development of S-101 represents a major step forward in product specifications for Electronic Navigational Charts (ENC). Based on the IHO geospatial framework standard S-100, S-101 will become the eventual replacement to S-57 ENCs. This paper will discuss the phased development approach that will lead to an ENC with improved functionality and better data handling throughout the ENC supply chain from producer to end-user, and will touch on several transition options that are under development.

1 INTRODUCTION

The International Hydrographic Organization (IHO) is an intergovernmental consultative and technical organization established in 1921 to support the safety of navigational, and to contribute to the protection of the marine environment. One of its primary roles is to establish and maintain appropriate standards to assist in the proper and efficient use of hydrographic data and information. (Ward et al. 2009)

This paper will describe the genesis of S-101 the new Electronic Navigational Chart Product Specification including its evolution from S-57 and S-52 – "Specifications for Chart Content and Display Aspects of ECDIS." The primary purpose of the paper is to communicate with the ENC user community the activities underway at the IHO in the development of this product specification and promote comment and the involvement from the maritime user community. The intent is not to develop S-101 in a vacuum, but to actively solicit input from software and equipment manufacturers and the ultimate end-user: the mariner.

1.1 S-57

One of the primary standards that the IHO is responsible for is IHO standard S-57, which is the current IHO "Transfer Standard for Digital Hydrographic Data." It was formally adopted in May 1992 and since that time it has become universally adopted as the underpinning standard for Electronic Navigational Charts (ENCs). S-57 Edition 3.1 was "frozen" in November 2000 and will remain so until no longer required. (Ward et al. 2009)

In January 2007, in response to an IMO update, the IHO released a supplement to S-57 to include new features and attributes required to encode Archipelagic Sea Lanes and Particularly Sensitive Sea Areas on ENCs. One of the characteristic features of S-57 is that the object and attribute catalogues defining the content of all ENCs is an integral part of the standard – thus a new supplement to S-57 was required to implement these new features.

According to an information paper published by the IHO, the following are current limitations of S57:
- It has an inflexible maintenance regime. Any addition of new features and attributes to the solitary catalogue for new products would have serious consequences for users of the ENC product specification such as ECDIS manufacturers, data production software vendors and regulatory authorities. It would trigger continual new editions because it freezes the object and attribute catalogues of the standard. Freezing the allowable content within data standards for lengthy periods is counter-productive for the end user.
- As presently structured, S-57 cannot support future requirements (e.g., gridded bathymetry, or complex time-varying information).
- Embedding the data model within the encapsulation (i.e., file format) restricts the flexibility and capability of using a wider range of transfer mechanisms while retaining data structure and content.
- It is regarded by some as a limited standard focused exclusively for the production and exchange of ENC data.

As a result of these limitations, in 2000 the IHO approved a major revision to S-57, resulting in the new framework geospatial standard S-100 (Ed 1.0.0 - January 2010).

1.2 *S-100*

S-100 provides a contemporary hydrographic geospatial data standard that can support a wide variety of hydrographic-related digital data sources, and is fully aligned with mainstream international geospatial standards, in particular the ISO 19000 series of geographic standards, thereby enabling the easier integration of hydrographic data and applications into geospatial solutions such as coastal zone management. S-100 extends the scope of the existing S-57 Hydrographic Transfer Standard. S-100 is inherently more flexible than S-57 and makes provision for such things as the use of imagery and gridded data types, enhanced metadata and multiple encoding formats. It also provides a more flexible and dynamic maintenance regime for objects, attributes and portrayal via a dedicated on-line registry.

The S-100 standard provides a framework of components that are based on, and designed to be interoperable with, the ISO 19000 series of standards and specifications. These standards and specifications are also used as the basis for most contemporary geospatial standards development activities and are closely aligned with other standards development initiatives such as the Open Geospatial Consortium (OGC).

The IHO has also developed an associated Registry to be used in conjunction with the S-100 standard. The IHO Registry contains the following additional components;
– Feature Concept Dictionary (FCD) Register.
– Portrayal Register.
– Metadata Register
– Register of data producer codes.

A particularly significant aspect of S-100 is that it provides the framework for the development of the next generation of ENC products, as well as other related digital products required by the hydrographic, maritime and GIS communities.(Ward, Alexander, and Greenslade, 2009) S-100 contains all the components necessary to create conformant product specifications to exchange a variety of digital hydrographic and marine geo-spatial information. S-100 contains multiple parts that were profiled from the ISO 19000 set of standards.

2 S-101

S-101 is the new Electronic Navigational Chart Product specification, currently under development by the IHO TSMAD Working Group. S-101 draws upon the concepts of S-100 such as exchangeable and dynamic feature and portrayal catalogues, and richer geometric models, information types and complex attributes. The use of these new feature types will allow ENC producers to overcome a number of known encoding shortcomings in S-57 based ENCs such as the overuse of caution areas,. In addition, the improved functionality will lead to more efficient data handling and better portrayal definition within ECDIS equipment, by eliminating or reducing the number of conditional symbology procedures.

One of the major benefits in S-101 is the ability to introduce additional functionality that is not available in S-57 ENCs. S-101 ENCs will eventually be the base navigation layer within an S-100 enabled ECDIS, but the true potential will not be realized until additional product specifications are developed to interact with S-101. Currently, the IHO has approved work for S-100 based product specifications for high resolution bathymetry and for nautical publications. Other potential S-100 based product specifications may include real-time tidal information and port operations information. A system capable of handling multiple S-100 products will allow for better navigational decision making by containing information regarding real time tides and sailing directions.

The key to the successful development of S-101 is the involvement of the maritime stakeholder community. This list includes: Hydrographic Offices, production software manufacturers, ECDIS equipment manufacturers, end-users, such as mariners, port authorities, and any other interested parties. During this process both TSMAD and the IHO are engaging in a continuous dialogue with stakeholders. In each of the past two years the IHO has held S-101 stakeholder workshops in order to present the status of S-101, the phased development process, and to receive feedback and suggestions on how best to overcome the current limitations in S-57 ENCs and the associated ECDIS portrayal requirements.

2.1 *Dynamic ENC Content*

The biggest advantage S-101 will have over the existing S-57 ENC product specification is the introduction of dynamic feature and portrayal catalogues. The term dynamic is used to characterize continuous change. While similar in content to the current S-57 object catalogue and the S-52 presentation library, S-101 implements the dynamic constructs prescribed by S-100. In S-101, the relationship between features, attributes and enumerants are defined within a single feature catalogue. Although, part of the standard, the feature catalogue is built through a registry responsible for defining data content and is machine readable, thus allowing ECDIS to easily update onboard systems via a software update. Under the current S-57 ENC regime updates to feature content may take up to five years to implement through the existing supplement process. Under S-100 the content of the registry is continuously changing, but the S-101 Feature and Portrayal Catalogues will be versioned, enabling the IHO to take advantage of the dynamic register content, but implementing a controlled update process for the end-users.

S-101 also defines a dynamic portrayal catalogue. This catalogue will replace the S-52 standard presentation library. The portrayal catalogue is a machine readable file containing IHO approved symbology and look-up table instructions to properly render ENCs on an ECDIS. This and the machine readable feature catalogue will make S-100 based ECDIS systems truly "plug and play".

S-101 will also make use of two new S-100 features to enhance the encoding, transfer, and portrayal of data. The first is a complex attribute, which has similar characteristics to the ISO 19000 *attribute of attribute*. A complex attribute is an aggregation of other attributes, either simple or complex (Figure 1).

The second is the use of "information types". An information type does not have any spatial attribution and will provide information about a feature by association. This can be used to represent a note associated with a pipeline or a buoy, for example. Under S-57, chart notes are typically encoded as a Caution Area – which is an alarm feature in ECDIS. Many of these notes contain relevant information and the only way to convey that information is through a Caution Area, however, most of the time this information does not need to signal an alarm. The creation of additional information type features will help reduce the amount of caution areas, a known encoding limitation within the current standard.

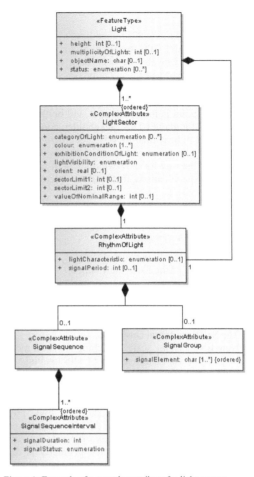

Figure 1: Example of a complex attribute for light sectors

3 THE PHASED APPROACH TO S-101 DEVELOPMENT

In order to address the concerns of stakeholders, at TSMAD 19 in October 2009, the membership agreed to a four-phase approach in the development, testing, and release of S-101. This approach will follow the methodologies of System Development Lifecycle Design.

The key benefits of this approach are: it allows for iterative development, where each iteration is usable and testable. It also allows for controlled and manageable change and does not try to solve everything at once. A phased approach will enable stakeholders to plan their development implementation of S-100 and will allow the hydrographic community to engage external stakeholders such as type approval authorities in the new standard.

3.1 *Phase 1: S-57 Content Equivalent*

Phase one of the development of S-101 is the "S-57 content equivalent". During this phase the existing feature content of S-57 is being replicated in an S-101 based product, using S-100 as its underlying framework.

The constraints for phase 1 are:
– The S-101 XML Feature Catalogue will be limited to only those features and attributes that are currently in S-57
– Introduction of S-100 geometry and the utilization of compound curves
– Use of the modified S-100 8211 encoding
– Use of complex attributes for light sectors and structured text attributes.

The major deliverables for this phase are an initial product specification and a content equivalent feature catalogue. The following benefits will be realized from phase one development:
– Proof of Concept and validation of S-100,

- Feature Catalogue can be exchanged as data – not specification,
- Utilization of a new 8211 encoding,
- Creation of an S-100 compliant product specification (which can also be used as a template for other S-100 product specifications),
- Proof that an S-57 based product can be built using S-100 mechanisms.

The last point is important as it will be the proof of concept for the ECDIS stakeholder community that S-101 will be similar to S-57, yet utilize the new functionality that is provided in S-100. In doing so it will pave the way for the development of additional product specifications that will be interoperable with S-101.

3.2 Phase 2: Enhanced Packaging and Data Loading Mechanisms

Phase 2 of S-101 will include
- the addition of the S-101 new support file formats and management,
- improved discovery metadata, from which ECDIS manufactures can inform the mariner which notice the dataset is corrected through.
- The S-101 portrayal catalogue
- New ENC display scales matching standard radar ranges that will be utilized for ECDIS loading.

During this phase TSMAD will also investigate the possibility of introducing scale independent and scale dependent data. If agreed by both TSMAD and the associated ECDIS stakeholders, ENC producers will be able to make the decision to partition a set of navigational data into two separate datasets based on whether their associated geometry is dependent on the compilation scale of the chart or not. The primary advantage of this structure is that receiving systems only need hold the scale independent features once, whereas in the current model multiple occurrences of features are required for different display scales. This in turn effectively reduces the size of the ENC dataset and increases the speed at which updates can be applied to datasets.

The deliverables for phase 2 will be an updated S-101 product specification addressing metadata and support file functionality, a revised feature catalogue and the first portrayal catalogue – effectively a translation of the current S-52 standard. The anticipated benefits from phase 2 are:
- Improved data delivery
- Improved data discovery
- Easily Accessible Metadata
- Comprehensive support for auxiliary file formats (xhtml, jpeg)

3.3 Phase 3: Extending the Model

During phase 3 the following will be included:

- More complex attributes introduced into the feature catalogue as well as the introduction of information types.
- Support for multiple languages
- Cartographic attributes that will support text placement, similar to a paper chart and alleviating the current cluttered text display associated with S-57/S-52 portrayal
- Revised guidance on pick reports enhancing better end user experience.

The updated feature and portrayal catalogues will allow for testing of the dynamic nature of S-101. The expected benefits from this phase are:
- Prove dynamic updating of feature and portrayal catalogues
- Enhanced language support

3.4 Phase 4: Scalability and Finalization

Phase 4 is the final phase in the development of S-101. At this point the first version of the product specification, feature catalogues and portrayal catalogues will be ready for approval by the IHO member states - probably in 2013. A tentative schedule for the entire development process is listed in Table 1.

Table 1. Tentative S-101 development schedule

Phase	Start Date	End Date
	MM/YYYY	MM/YYYY
Phase 1	01/2009	12/2010
S-57 convertor	11/2010	04/2011
Phase 2	01/2011	08/2011
Phase 3	09/2011	03/2012
Phase 4	04/2012	12/2012

4 TEST BEDS

Another important element in the development of the S-101 product specification is the requirement for test beds during the development lifecycle and beyond. TSMAD has begun the process of identifying items needed for the test beds. The main items are as follows:
- S-57 to S-101 open-source convertor
- S-101 open source data editor
- S-101 open source data viewer
- S-100/101 ECDIS reference Test Bed

In recognizing the need for test beds and to help promote the development of the S-101 Product Specification, the National Oceanic and Atmospheric Administration (NOAA) contracted with ESRI to develop an S-57 to S-101 open source convertor. Once completed, NOAA will turn this over to the IHO to be placed in the public domain. This convertor is intended to convert existing S-57 ENC data into S-101 ENC data by utilizing the feature catalogue developed in phase one of the project plan. It will also utilize the new ISO8211 encoding and provide samples of S-101 test data for interested stakeholders. It is expected that this convertor will be completed in March 2011.

TSMAD has also recognized the need for both an S-101 data editor and a viewer to enable the creation of S-101 data from scratch. This is so the functionality of the exchangeable feature and portrayal catalogues can be proved and to allow the creation of test data using new S-101 functionality.

The final test bed required will, in effect, be a reference S-100 ECDIS. It will enable TSMAD to test the updateable feature and portrayal catalogues in an environment that is modeled on genuine displays suitable for type approval.

5 TRANSITION

Although, S-101 is not expected to be adopted by the IHO as a standard before at least 2013, it is not too early to begin the process of informing the maritime community of the development of S-101. That is the purpose of this paper.

One of the major issues that will need to be agreed by the IHO and relevant stakeholders is how to transition from S-57 ENCs to S-101 ENCs. Obviously current S-57 enabled ECDIS will not be able to view S-101 data and there will have to be a lengthy transition period to enable both end-users, manufacturers, ENC producing countries and regulatory authorities to changeover. The successful development of S-57 to S-101 data converters will likely play a key part in determining how quickly the changeover could be made.

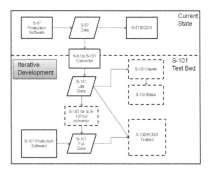

Figure 2: S-101 Test Bed

REFERENCES

Ward et al. 2009. IHO S-100: The New IHO Hydrographic Geospatial Standard for Marine Data and Information. International Hydrographic Organization. Monaco

S-100 IHO Universal Hydrographic Data Model (Ed 1.0.0 – January 2010). International Hydrographic Organization. Monaco

S-57 IHO Transfer Standard for Digital Hydrographic Data, Edition 3.1 November 2000. International Hydrographic Organization. Monaco

ISO/IEC 8211:1994: Information technology – Specification for a data descriptive file for information interchange. International Organization for Standardization. Geneva, Switzerland.

ISO 19100 Series – Geographic information/Geomatics. International Organization for Standardization. Geneva, Switzerland.

Visualization and Presentation of Navigational Information

10. Applications and Benefits for the Development of Cartographic 3D Visualization Systems in support of Maritime Safety

R. Goralski
GeoVS Ltd, Cardiff, UK

C. Ray
Institut de Recherche de L'Ecole navale

C. Gold
Faculty of Advanced Technology, University of Glamorgan, UK

ABSTRACT: Maritime shipping is among the world's most important industries and is vital to the global economy. With growing levels of traffic marine accidents pose a danger to health and lives of ship crews, environment, and have a strong impact on profitability of shipping and port operations. This underlines the urgent need for the development of maritime navigation systems whose objective will be to contribute to a safer sea. Amongst many technical and methodological issues to address, there is a need for more efficient electronic charting and radar display systems for use in navigation, traffic monitoring and pilotage, to improve the level of situational awareness of ship navigators, Vessel Traffic Services (VTS) operators and marine pilots. Cartographic 3-dimensional visualization (3D chart) facilitates fast and accurate understanding of navigational situations in ports and at open sea, decreases mental overload and minimises fatigue, this supporting better decisions for either sailors at sea or maritime authorities in charge of traffic monitoring. This leads to a reduction of human error which is the main cause of marine accidents. This paper presents the latest developments and applications of cartographic 3D visualizations (3D charting) in marine navigation, VTS and pilotage.

1 INTRODUCTION

Maritime shipping is among the world's most important industries. It can be compared to a cardiovascular system of the global economy – for thousands of years ships of different types and sizes are transporting goods between different countries and continents, facilitating world trade by enabling some routes and making other economically viable.

The risk of marine accidents has always be a major consideration, as a factor which not only determines the economic viability and profitability of shipping, but also directly endangers health and lives of ship crews. Moreover, ships carrying hazardous loads pose serious threats to the environment, as well as to the lives, health and wellbeing of the people and animals inhabiting coastal zones. The disaster and damage caused in the event of a major sea collision can be difficult and costly to deal with. Collisions of even small craft often lead to serious consequences, often including casualties.

Over the years huge efforts have been directed towards improving the level of maritime safety, including areas as diverse as ship design, certification, training, fire safety, radio-communications, navigation rules, electronic chart systems, identification systems, life-saving equipment, ship operation and maintenance procedures and requirements, port operations, pilotage and accident investigation. International treaties such as the International Convention for the Safety of Life at Sea (SOLAS) (International Maritime Organisation 2004) have been introduced to prevent accidents by regulating those and other maritime safety-related areas. A huge progress has been achieved.

However, with the level of shipping constantly growing in the long term, and reaching its historical peak levels in the late XX and early XXI centuries, the risks of accidents from categories as diverse as equipment faults and breakdowns, fires and explosions, personal injuries, collisions and groundings are also growing, with the last two categories responsible for the majority of fatalities (Talley et al. 2006).

According to research by the U.S. Coast Guard Research & Development Center (Rothblum 2006) about 75-96% of marine casualties are caused, at least in part, by some form of human error, with this being the case for 84-88% of tanker accidents, 79% of towing vessel groundings, 89-96% of collisions and 75% of fires and explosions. Those numbers show clearly that human error ranks as the major contributor to different types of marine accidents, including the most dangerous categories such as collisions and groundings.

As shipping lanes are increasingly becoming crowded with larger and faster craft, ship crews are getting smaller. This puts a growing pressure on the navigators, as well as on the on-board systems that are required to provide them with decision-making and navigation support.

Human errors can also come from control centre (Vessel Traffic Services, VTS) on shore where people might have difficulties to appreciate and anticipate a given situation due to the traffic load. Also marine pilots, despite their excellent knowledge of navigation and local expertise are not immune to errors. Investigation of accidents such as the grounding of Vallermosa (Marine Accident Investigation Branch 2009) indicate insufficient level of support from ship crews, mental overload, inability to comprehend and control the developing scenario, the lack of situational awareness, and the absence of coordination and support from VTS operators among main causes of accidents involving pilot-led vessels.

The latest technological breakthroughs including radar, electronic charting (Electronic Chart Display Information Systems, ECDIS), traffic control and management (VTS) and automatic identification and communication (Automatic Identification System, AIS) brought a significant improvement to the problem of maritime navigation safety, contributing greatly to improved navigational awareness, collision-avoidance information and guidance available to navigators. However, they have not eradicated the problem and marine accidents still happen frequently, often due to fatigue, mental overload and limited awareness of the navigational situation.

This can be improved by offering more visually efficient and both easier and quicker to understand chart display systems based on cartographic 3D visualization to navigators, pilots and VTS operators. Three-dimensional charts are proven to dramatically reduce the number of human mistakes and improve the accuracy and time efficiency of navigational operations, compared to traditional 2D charts (Porathe 2006), including Electronic Chart Display Information Systems (ECDIS). To minimise human error, and in consequence reduce the number of accidents, they could be applied at several stages of the maritime safety management process, including on-board ship navigation, vessel traffic monitoring (VTS) and pilotage. They also can be used in training and provide significant insight in accident investigation.

2 NAUTICAL ELECTRONIC CHARTS

Maps are among the oldest forms of graphical communication, and are the most effective and efficient mean for transfer of spatial and geographical information (Kraak 2001). Over the years, different types of maps have been developed with different cultural and application backgrounds, and used for many aspects of our everyday lives. More recently, maps have moved from paper to digital formats and are becoming more popular than ever.

From the navigational perspective electronic charts offer a number of benefits over their paper equivalents, allowing for dynamic analysis of vessel position and chart data for alerting about potential groundings, integration with bridge equipment and presentation of combined information from sensors such as GPS, radar and AIS, and automation of typical navigational tasks, such as plotting the course or calculating different parameters of the planned route. This integration and automation helps in reduction of navigators' workload, and offers more accurate understanding of the navigational situations.

2.1 State-of-the-art

Due to its obvious benefits electronic charting is fully supported, and encouraged, by the International Maritime Organization (IMO), International Hydrographic Organization (IHO) and member state regulators, who developed a standard for Electronic Chart Display Information System (ECDIS), with an objective to replace maritime maps on the decks of commercial ships with automated and electronic charts. An ECDIS system comprises the official nautical chart data (International Hydrographic Bureau 2000) stored in the vector IMO/IHO Electronic Navigational Chart (ENC) format produced by national hydrographic offices (Fig. 1), a type-approved real-time 2-dimensional display conforming to performance and display standards associated with the current position of a vessel obtained from GPS, a user interface to perform basic navigational tasks, with optionally integrated information from AIS, radar, and other bridge instruments.

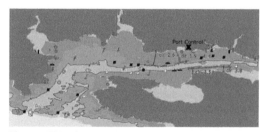

Figure 1. ENC chart no GB50162B – the Port of Milford Haven – as viewed in an ECDIS, with additionally marked location of the Port Control (VTS operations centre)

The carriage of type-approved ECDIS will become mandatory on all merchant and passenger ships with a transitional schedule for the implementation of this requirement for different types of new and existing ships starting from July 2012 (new passenger ships of 500 gross tons or more, and new tankers above 3000 gt) to July 2018 (retrofit on existing dry cargo ships of above 10000 gt).

VTS centres and pilots usually use non-approved chart display systems based on official ENC charts (Electronic Chart Systems, ECS), offering special functionality for situation analysis and tracking of port operations.

Leisure boating enthusiasts have a broad selection of chart plotters, which do not have to conform to regulations, available on the market, and these are constantly gaining in functionality and popularity.

2.2 *Supporting technologies*

The main purpose of a charting system, or an EC-DIS, is to display the relevant chart and navigational information and to automatically present the ship in this context. This is done by overlaying the ship's position as received from a satellite navigation transponder on a digital chart.

However, one of the most appealing functionalities of digital charts is the possibility of easy integration of additional information received from other systems or on-board sensors and devices. This may include meteorological data (from weather station, or from weather forecasts), digital compass, and other, but most importantly the information about surrounding obstacles and traffic which is very helpful in avoidance of dangers. This information may come from radar (drying features, obstacles, other ships), sonar (bathymetry), but also from Automatic Identification System (AIS) which is one of the most significant recent technological developments, allowing for reliable and easy detection and identification of ships within a range of up to 35 NM. An AIS transponder generally integrates a transceiver system, a GPS receiver and other navigational sensors on board such as a gyrocompass and a rate of turn indicator. It runs in an autonomous and continuous mode, regardless of its location (e.g., open sea, coastal or inland areas). The transmitted information is broadcasted through VHF communications, to surrounding ships and to VTS systems operated by maritime and port authorities.

The International Maritime Organization has made the AIS a mandatory standard under the Safety of Life at Sea (SOLAS) convention for all passenger and international commercial ships, and these are now equipped with AIS transponders.

2.3 *Limitations*

Despite the progress brought with the shift towards digital charting and introduction of ECDIS and other technologies, 2-dimensional maps are sometimes difficult to interpret from a cognitive point of view. Users have to generate a mental model of a map, rotate it and match with real world and translate the symbols and map features towards some abstract concepts. This explains why so many people have difficulties with the interpretation and understanding of 2-dimensional maps, this often resulting in errors and sometimes leading to fatal mistakes.

This is true even for trained navigators, especially when they are tired or under pressure, with high levels of stress and mental overload, which are typical for marine navigation. The time and mental effort required for understanding of 2-dimensional maps has severe consequences in areas where time for analysis of the situation is crucial. This is the case for example in navigation of high-speed marine craft, where not only situation changes quickly, and available time for reaction is limited, but tiredness of navigators further limits their cognitive capabilities. (Porathe 2006) describes several cases of high speed vessels crashing into rocks, among them one involving 16 death casualties. All these collisions were caused by poor understanding of the situation or temporal loss of orientation by navigators, despite being guided by modern digital 2D chart displays.

3 3D VISUALIZATION IN MARITIME SAFETY

3.1 *State-of-the-art*

The application of 3D visualization to maritime safety is not a new idea. 3D presentations are widely adopted, extensively used and highly regarded in some areas of the maritime safety management process, such as for example in marine navigation training, where realistic 3D simulators are a safe, cheap, convenient and reliable way of gaining navigational experience, or in ship building, where three-dimensional Computer Aided Design (CAD) systems are extensively used for design and modelling.

However, for some reasons, 3D visualization has not been successfully adopted in real-time 3D navigation, VTS or pilotage. There are no professional 3D charting systems available on the market, with very limited success of adoption of 3D perspectives in chart plotters offered to the leisure market. The same situation is true in the VTS or pilotage operations where ECDIS-like 2D chart displays are commonly used. The possible reasons for this are discussed in Section 4.2.

3.2 *Cartographic vs. photorealistic 3D*

It is important to stress the importance of the distinction between photorealistic and cartographic 3D presentations. The representatives of the first group are meant for realistic representation, or mimicking of the real world, and are known from computer games and navigation training simulators. The goal in simulation is clear: to recreate the situation and conditions of navigating at sea with greatest possible accuracy, to offer a trainee as much practice time as possible in diverse nautical conditions, without the risk, time and cost of going into the sea. As such, simulators have to be built to represent the realism of the situation with all its negative aspects, such as inability to see more than the view out of the window and of what typical bridge instruments would present, which offers a limited situational awareness.

Just as a well-designed 2D chart differs from a photograph, cartographic 3D presentations are different from their photorealistic counterparts. Cartographic 3D visualizations (3D maps) enhance and facilitate the understanding of the presented situation, by clarifying and tailoring the presentation to user needs, and are designed for the highest possible efficiency of information transfer. As outlined in more detail in Section 5.1, to achieve that purpose they use cartographic principles equivalent to or extrapolated from 2D cartographic rules.

3.3 Potential and benefits

Despite the observation that 3D visualization is currently not present at all or used only to a very limited degree in maritime navigation, research and experiments indicate that the application of 3D in marine charting may offer significant benefits, including faster and more accurate understanding of the portrayed situations and higher level of operational comfort, when compared to their 2D counterparts.

This potential is based on the inherent ease of understanding of 3D representations which is a consequence of the way we see the world, and how 3-dimensional representations appeal to our brains (Van Driel 1989). The process of perception in three dimensions has been perfected by millions of years of evolution, because prompt recognition of potential dangers was, and still is, crucial for survival. According to estimates, about 50% of brain neurons are used in the process of human vision. 3-dimensional views stimulate more neurons and hence are processed quicker (Musliman et al. 2006). 3-dimensional maps resemble the real world to a greater degree than their traditional 2D counterparts, and are more natural to human brain (Schilling et al., 2003). Another advantage is that 3D symbols can be recognized very quickly even without special training or referring to a legend.

Based on experiments conducted with different types of maps, Porathe (2006) argues that 3-dimensional maps are not only quicker to understand but also provide improved situation awareness, and have strong potential of helping to minimize human error in marine navigation and the resulting marine accidents. In his experiments Porathe asked a group of participants with different characteristics (age, sex, navigational experience) to perform a simulated navigational exercise using four different types of charts: paper, digital 2D north-up, digital 2D heads-up and interactive 3D chart. The efficiency of navigation using each map type was measured as the time required to complete the task, the number of mistakes (groundings) made during its execution, and the perceived difficulty of use. The results showed that the use of 3-dimensional maps led to

up-to 80% reduction in the number of navigational mistakes, as well as more than 50% reduction in the time required for the completion of the simulated navigational task, when compared to 2D charts. Three-dimensional maps were also voted by the participants as the most friendly and easy to understand.

A very important finding was that the patterns of the results were similar in every participants group, regardless of their navigational experience level. While on average experienced navigators completed their tasks faster and caused less groundings a similar increase of efficiency and reduction of error was observed for their group, as for inexperienced users. And just as inexperienced group trained navigators appreciated the perceived friendliness of and ease of navigation using 3D charts.

It is our belief that this benefits can and should be transferred from the experimental research domain into the world of real navigation, to offer the benefits of 3D charting to navigators, and reduce the number of accidents in ports and at sea. The improvements to maritime navigational safety can be gained by application of 3D visualization to navigational charts for use in real-time in navigation, VTS operations and pilotage, or for analysis, training and accidents investigation with use of historical data recordings.

4 NAUTICAL 3D VISUALIZATION

4.1 Previous work

The idea of 3-dimensional navigational charts was initially introduced in (Ford 2002) with the conclusion that 3-dimensional visualization of chart data had the potential to be a decision support tool for reducing vessel navigational risks. A prototype charting system was based on custom-prepared 3D model of the selected area.

Figure 2. Bridge view in the "3D ECDIS" prototype

Arsenault et al. (2003) presented a prototype 3D visualization system that used an overlay of a scanned paper navigational chart over a 3D bathymetry, and served as a platform for research on concepts that might be used in the "chart-of-the-future," including merging tidal and bathymetric information and simultaneous display of multiple linked views.

Gold et al. (2004) proposed a prototype of an interactive 3-dimensional "pilot-book" for the East Lamma Channel in Hong Kong, with 3D model (map) of the area derived from the corresponding ENC cell, and manually converted into a 3D presentation with the use of satellite DTM and ortho-imagery.

Porathe (2006) introduced a prototype charting system, with custom-built visual vocabulary of guidance symbols, comparable to virtual signs on a motorway, for use in navigation. The system worked on manually prepared 3D model of the selected area.

Goralski & Gold (2008) proposed a "3D ECDIS" prototype based on a custom-designed 3D visualization engine, kinetic spatial data structures and ergonomic manipulation interface (Fig. 2). The system used official ENC charts to automatically create real-time 3-dimensional display associated with the current position of vessels.

Figure 3. Immersive real-time 3D visualization of a sailing regatta (Brest Bay)

Ray et al. (2011) presented a 3D virtual environment based on an interactive 3D chart for tracking of marine vessels and visualization of a sailing regatta competition for a wider public in real-time (Fig. 3). The system was based on a real-time tracking and dissemination platform introduced by Bertrand et al. (2007, Fig. 4).

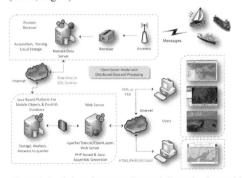

Figure 4. Real-time data recording and dissemination architecture

Ternes et al. (2008) proposed a prototype 3D visualization system developed with the Port of Melbourne, for support of navigation during hydrographic surveys. System works with manually-prepared 3D models and proved to be very effective in helping to maintain accurate survey track with the use of virtual boys and markers.

4.2 Inhibiting factors

After analysing the benefits of 3D visualization to maritime navigation safety, including the results of experiments, and reading about various prototype developments proposed in the last decade an important question occurs. If 3D charts have been proven to be so efficient and have been proposed by a number of researchers, why they have not become popular and are not used widely for navigation? Why they seem to never have moved beyond research prototypes stages? And why, unlike hugely popular 3D marine navigation simulators, they are not commercially successful and available on the market? These are complex questions, and our experience suggests that there are several combined reasons for this situation.

The first reason is in technical complexity. Building a robust, reliable and usable 3D charting product that is designated for constant real-time use in a very important and responsible function, upon which the safety of people relies (which an aid to navigation certainly is) is a much more demanding task than is the case for either digital 2D chart or a training simulator.

This is combined with well-meant and fully justified conservatism and scepticism of the regulators and the industry. It took years for the industry to acknowledge and appreciate the benefits of 2D charting and to develop and embrace standards such as ECDIS. The regulators, shipping operators and navigators have to be cautious in entrusting any new technologies which have not been fully validated and tested in practice.

The technical complexity and industry conservatism make development of professional 3D charting a costly and relatively risky endeavour, which requires more resources and a longer development time, with significantly more difficult quality control, while not necessarily leading to gains which would compensate for it.

On top of that, but partially resulting from the above problems, are the legal limitations. We cannot really have a 3D ECDIS at the moment, because it would not fulfil very strict and precise standards of information presentation set out by the regulators. 3D is not an option for type-approved ECDIS displays and it seems unlikely that this will change for many years to come.

Another consideration is the cost of data acquisition and the availability and coverage of 3D charts.

In all prototype systems listed above, with the exception of the prototype "3D ECDIS" by Goralski & Gold (2008), 3D charts had to be manually created in a laborious process, which restricted their usability to the selected areas of interest.

Apart from the other reasons, there is the conceptual difficulty and lack of experience, tradition and know-how in efficient presentation of cartographic information in 3D. Unlike 2D cartography which has been practised and developed for hundreds or even thousands of years, 3D cartography is rather new and is a largely uncharted territory. Several researchers in the area, including Haeberling (2002) and Meng (2003) complained about the lack of available research in the field, and the situation has not improved since. That leads to a situation where very little knowledge and guidance is available to potential producers of 3D mapping systems.

Part of the above refers to the difficulty of building efficient presentations of data in 3D, while another part concerns the complexity of design of ergonomic user interfaces for efficient operation of (or navigation within) 3D charts, which with the addition of the extra dimension becomes an incomparably more difficult task than in 2D.

All the above factors combined contribute to the delay in adoption of 3D charting, but in our opinion none of the discussed reasons should stop this beneficial process, providing that the requirements for successful 3D charting products will be fully understood, and the identified difficulties solved.

4.3 *Requirements*

From the analysis of the difficulties inhibiting and delaying the popularisation of 3D charting in marine navigation it seems clear that successful 3D charting products should fulfil a number of requirements.

Firstly, they should be at least as robust and reliable as their 2D counterparts are, to be able to overcome the reservation and scepticism with which the regulators and maritime industry rightly treat all new technologies.

Secondly, 3D charts should be applied first where legal regulations do not preclude them. There are no reasons why they could not be used in VTS or pilotage, and in professional navigation they can be used as add-ons to, or alongside, rather than instead of, type-approved ECDIS.

Thirdly, to tackle the problem of costly data acquisition and availability they need to be universal. Ideally they should work with existing, approved and widely available standards of chart data, such as ENC charts, and generate 3D models of the covered areas automatically.

Fourthly, they need to be user-friendly and ergonomic. Their operation and manipulation has to be effortless, natural and intuitive, and cannot be cumbersome or distracting.

Finally, they have to use cartographic principles to maximise the efficiency of information presentation and transfer.

Fulfilling all the above requirements may require extra cost and time for research and development, but is necessary to produce 3D charts that can be used and relied upon even in the most demanding conditions, and should be worthwhile due to benefits that well-designed 3D charts would bring to marine navigation.

5 NAUTICAL 3D CHARTS

5.1 *3D Cartography*

Cartography is the discipline dedicated to study and practice of production, interpretation and use of maps, and includes all related scientific, technical and artistic activities and aspects (Edson 1979). The expertise of cartographers involves the knowledge of visual design – to make maps as readable as possible – as well as diverse related areas such as the process of map production and distribution, different forms of map use and applications but also the underlying technologies and algorithms, as well as the psychological mechanisms of human perception and cognition.

The scope of 3D cartographic knowledge is an extension of the traditional cartography, but includes areas and aspects which either need to be adapted for the use with the additional dimension, or are completely new.

Visual design is one of the examples from the first category. Cartographic presentations, i.e. maps, require carefully designed symbols and methods for presenting different types of information. They use the principles of symbolism, generalisation and abstraction. In 3D some symbols, texts and numbers may be presented using methods similar to those known from 2D. Other may need to be represented differently – for example as self-explanatory 3D models of the real-world objects, including the 3D model of the terrain (and bathymetry) of the presented area. The 3D representations do not have to be photo-realistic to be easily understandable. In fact non-photorealistic computer graphics is known for its capability to provide vivid, expressive and comprehensive visualizations with strong potential for use in cartography (Durand 2002, Dollner 2007). Desirable are dynamic algorithms for the optimisation of the chart display, to assure that the important information is always presented efficiently.

Interactivity and manipulation interfaces are among the examples of aspects which are completely new, or incomparably more important and complex in 3D. Static 3D (perspective) presentations have been known for years, but never gained much popularity. This is due to their inherent limitations, including the distortion of the perspective and the problem of hidden regions. It is with the introduction of the ability to interact with or within the presented environments when 3D maps become truly beneficial. A user of a 3D map needs to be able to freely move in the represented area and test different perspectives to be able to fully understand the presented situation. For that reason the design of highly interactive and ergonomic interfaces emerges as a new area of interest for 3D cartography. The use of typical "arrows" (pan left, right, up and down) and "zooming" (zoom in and out) buttons in a 3D map is not a satisfactory option. Preferred are direct manipulation interfaces which allow for constant and fluent control of all the required levels-of-freedom in an ergonomic, effortless and intuitive manner.

The trouble with 3D cartography is the already mentioned lack of sufficient research results, resources and know-how which would guide producers in their efforts of the development of truly usable and efficient 3D charting products.

5.2 State-of-the-art

The deficiency of 3D cartographic research means that each 3D chart development effort has to be based not only on the existing body of knowledge, but also add a significant amount of original work into the subject.

This paper proposes a 3D charting (cartographic 3D visualization) system for marine navigation, VTS and pilotage, using best practices and the scarce research available on 3D cartography, as well as our experiments conducted and experience gathered during over the decade of its development at the Hong Kong Polytechnic University, University of Glamorgan, French Naval Academy and GeoVS Limited. The system was designed based on the requirements described in Section 4.3.

The navigation, VTS and pilotage systems are built around the common 3D charting platform "C-Vu" and are called respectively "C-Vu 3D ECDIS" (add-on to type-approved systems), "C-Vu Surveillance 3D VTS" and "C-Vu 3D Pilot" (for pilot boat navigation, as well as pilot carry-on units.)

Apart from highly efficient 3D cartographic engine the systems integrate with port and on-board infrastructure and sensors (GPS, AIS, radar, tide gauges, weather stations, digital compass, ship register, networking equipment) and offer robust recording and distributed information dissemination architecture for real-time remote multiple-display monitoring as well as analysis and evaluation of historical data – for training or accident investigation. Different elements of the system integrate with each other, allowing a pilot with a carry-on unit to see the complete picture of the situation in the port as recorded by the VTS, or for a mobile radar installed on a pilot boat to feed into the main system, thus increasing

the local clarity of the radar picture or covering blind spots. The architecture allows for redundancy of all system elements for increased reliability.

In terms of usability "C-Vu" 3D charts offer a global coverage with automatic generation of fully-fledged 3D models from official ENC charts, with the use of advanced Voronoi algorithms for terrain and slopes reconstruction (Dakowicz & Gold 2003). The user interface is kept simple and ergonomic – it is tailored for the purpose of each system version – and employs our original metaphor for navigation in the 3D map area, which is coupled with a hardware 3D controller to allow efficient direct manipulation in all directions with a single hand.

The 3D cartographic engine is built around a set of custom-designed 3D models of navigational objects and different ship types, with dynamic algorithms for optimisation of sizes, colours and spatial orientations, to assure the highest possible level of visual efficiency.

Underneath the visual layer are GIS-type kinetic Voronoi-based algorithms that maintain spatial relationships between different objects in the model (including the bathymetry), and may be used for prediction and avoidance of collisions and groundings.

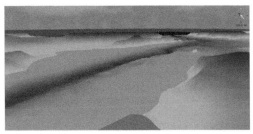

Figure 5. C-Vu Surveillance 3D VTS - bathymetry and terrain model of the Port of Milford Haven as seen from the location of the Port Control (traffic information not displayed)

All systems from the "C-Vu" family are in their final development stage. "C-Vu Surveillance 3D VTS" is currently being used and trialled for 24/7 VTS operations, and is being improved with the operators feedback, in the Port of Milford Haven in Wales, UK (Fig. 5). It is expected to be commercially available in the summer 2011 from GeoVS Limited, with "C-Vu 3D ECDIS" and "C-Vu 3D Pilot" following later that year.

5.3 Future work

Future work will include the finalisation of the development of the navigation and pilotage systems from the "C-Vu" family, improvements of the distributed platform, as well as continued research work on 3D cartography, visual efficiency, user-chart interaction, decision support, and 3D charts' applications.

Special focus will be placed on providing users with efficient decision support, through employment of the underlying spatial algorithms to the avoidance of accidents, and the use of advanced behavioural models for prediction of dangerous behaviour from the participants of marine traffic (Le Pors et al. 2009).

Once a sufficient user base is built a team of occupational psychology researchers from the University of Glamorgan will undertake a formal assessment, including qualification and quantification, of the occupational comfort and efficiency benefits attained by the use of 3D charts in VTS and navigation.

6 CONCLUSIONS

Although current maritime systems are relatively successful, thanks to the development of digital charting, vessel traffic systems and automated information and communication systems, there is still a need for innovation in minimising human error and reducing the number and severity of accidents at sea.

With advances in cognitive science and psychology comes a better recognition of the strengths and limitations of human perception and natural abilities of our brains. New charting products should use this knowledge and be designed to support the strengths and minimise the limitations, to reduce errors made by navigators, VTS operators and pilots.

3D charts were proposed and tested as a very efficient medium for fast and accurate transfer of navigational information and enhancement of situational awareness of ship crews, VTS operators and pilots, even when working under pressure or when fatigue significantly limits their cognitive capabilities. 3D charts reduce mental overload and improve overall occupational and operational comfort and efficiency.

The research presented in this paper explains the benefits of 3D visualization in maritime safety, presents the state-of-the-art of 3D charting and introduces a universal 3D chart display system based on official ENC charts and integration with bridge or port infrastructure.

The proposed system generates fully-fledged cartographic 3D models of any area from standard ENC charts and is being prepared for use in marine navigation, VTS and pilotage, where it provides a better sense of the environment for many situations at sea, this being an asset for safer navigation. The interface developed can be used either in real-time for navigation monitoring and control, or for the analysis of maritime navigation behaviours. The system is currently being trialled in the Port of Milford Haven for its VTS operations, and has a potential to become the first fully-functional commercial implementation of a 3D VTS.

REFERENCES

Arsenault, R., Plumlee, M., Smith, S., Ware, C., Brennan, R. & Mayer, L. 2003. Fusing Information in a 3D Chart of the Future Display. In *Proceedings of the U.S. Hydro 2003 Conference, Biloxi, MS, USA, 24-27 March 2003.*

Bertrand, F., Bouju, A., Claramunt, C., Devogele T. & Ray, C. 2007. Web architectures for monitoring and visualizing mobile objects in maritime contexts. In G. Taylor & M. Ware (eds.), *Proceedings of the 7th International Symposium on Web and Wireless Geographical Information Systems (W2GIS 2007), Cardiff, UK, November 2007. LNCS 4857*: 94-105. Springer-Verlag.

Dakowicz, M. & Gold, C.M. 2003. Extracting Meaningful Slopes from Terrain Contours. *International Journal of Computational Geometry and Applications* 13: 339-357.

Dollner, J. 2007. Non-Photorealistic 3D Geovisualization. In W. Cartwright, M.P. Peterson & G. Gartner (eds.), *Multimedia Cartography (2nd Edition)*: 229-239. Springer.

Durand, F. 2002. An Invitation to Discuss Computer Depiction. In *Proceedings of the 2nd international symposium on Non-Photorealistic Animation and Rendering (NPAR), Annecy, France.*

Edson, D.T. 1979. The International Cartographic Association - An Overview. In *Proceedings of the 4th International Symposium on Cartography and Computing: Applications in Health and Environment (Auto-Carto IV), Reston, Virginia, USA, 4-8 November 1979*: 164-167.

Ford, S.F. 2002. The first three-dimensional nautical chart. In D.J. Wright (ed.), *Undersea with GIS*: 117-38. ESRI Press.

Gold, C.M., Chau, M., Dzieszko, M. & Goralski, R. 2004. 3D geographic visualization: the Marine GIS. In P. Fisher (ed.), *Developments in Spatial Data Handling*: 17-28. Berlin: Springer.

Goralski, R. & Gold, C. 2008. Marine GIS: Progress in 3D visualization for dynamic GIS. In A. Ruas & C. Gold (eds.), *Headway in Spatial Data Handling*: 401-416. Springer.

Haeberling, C. 2002. 3D Map Presentation: A Systematic Evaluation of Important Graphic Aspects. In *Proceedings of the 3rd ICA Mountain Cartography Workshop, Mt. Hood, Oregon, USA, 15-19 May 2002.*

International Hydrographic Bureau 2000. *IHO transfer standard for digital hydrographic data edition 3.0, Special publication No. 57.*

International Maritime Organization 2004. *SOLAS: International Convention for the Safety of Life at Sea, 1974 (Consolidated Edition 2004).* IMO Publishing.

Kraak, M.J. 2001. Cartographic principles. In M.J. Kraak & A. Brown (eds.), *Web Cartography: Developments and Prospects*: 53-72. CRC Press.

Le Pors, T., Devogele, T. & Chauvin C. 2009. Multi-agent system integrating naturalistic decision roles: application to maritime traffic. In *Proceedings of the IADIS 2009, Barcelona, Spain, 25-27 February 2009.*

Marine Accident Investigation Branch (2009). *Vallermosa, Investigation Report No 23/2009.*

Meng, L. 2003. Missing Theories and Methods in Digital Cartography. In *Proceedings of the 21st International Cartographic Conference, Durban, South Africa.*

Musliman, I.A., Abdul-Rahman, A. & Coors, V. 2006. 3D Navigation for 3D-GIS – Initial Requirements. *Innovations in 3D Geo Information Systems*: 259-268. Springer-Verlag.

Porathe, T. 2006. *3-D Nautical Charts and Safe Navigation, Doctoral Dissertation.* Mälardalen University Press

Ray, C., Goralski, R., Claramunt, C. & Gold, C. 2011. Real-time 3D monitoring of marine navigation. In *Proceedings of the 5th International Workshop on Information Fusion and Geographical Information Systems: Towards the Digital Ocean (IF&GIS 2011), Brest, France, 10-11 May 2011.*

Lecture Notes in the Geoinformation & Cartography. Springer-Verlag.

Rothblum, A.M. 2006. Human Error and Marine Safety. *U.S. Coast Guard Risk-Based Decision-Making Guidelines vol. 4.* U.S. Coast Guard Research and Development Center.

Schilling, A., Coors, V., Giersich, M. & Aasgaard, R. 2003. Introducing 3D GIS for the Mobile Community - Technical Aspects in the Case of TellMaris. In *Proceedings to the IMC Workshop on Assistance, Mobility, Applications in Stuttgart.*

Talley, W.K., Jin, D. & Kite-Powell, H. 2006. Determinants of the Severity of Passenger Vessel Accidents. *Journal of Maritime Policy & Management*, 33(2), 173-186.

Ternes, A., Knight, P., Moore, A. & Regenbrecht, H. 2008. A User-defined Virtual Reality Chart for Track Control Navigation and Hydrographic Data Acquisition. In A. Moore & I. Drecki (eds.), *Geospatial vision, new dimensions in cartography: selected papers from the 4th National Cartographic Conference GeoCart' 2008, New Zealand*: 19-44. Springer.

Van Driel, N.J. 1989. Three dimensional display of geologic data. *Three Dimensional Applications in Geographical Information Systems*: 1-9. CRC Press.

11. Assumptions to the Selective System of Navigational-maneuvering Information Presentation

R. Gralak
Institute of Marine Traffic Engineering, Maritime University of Szczecin, Poland

ABSTRACT: In the era of emerging technologies in the transport decided to create three-dimensional visualization system which virtualizes real navigation situation of the ship in a restricted area. The system in its destiny is a part of a large branch of the eNavigation and is intended as a tool to assist decision navigator on the ship's bridge, particularly in the berthing maneuvers. The article presents the technical assumptions for the system. Presents its destination, innovative solutions including the ability to multi-territorial virtualization and preview the actual position of the individual.

1 INTRODUCTION

For several years in the maritime field, we distinguish the "eNavigation" term. The IMO's eNavigation initiative has as its goal the seamless integration of information: "eNavigation is the harmonized collection, integration, exchange, presentation and analysis of maritime information onboard and ashore by electronic means to enhance berth to berth navigation and related services, for safety and security at sea and protection of the marine environment" (IMO, 2004).

The decision support systems (Navigation Aids) also enroll into this area, which covers broad scope of information to assist the navigator in the safe passage of the ship. These include:
- classified hydrometeorological information,
- anti-collision and alarm systems,
- cargo handling operations systems,
- systems to monitor ship traffic parameters,
- for the route monitoring systems,
- systems for presentation of vessel's location in space
- other.

We also note the tendency to integrate multiple systems, with the properly selected, multi-level form of presentation of navigational information - Human Machine Interface (HMI), forming the Integrated Bridge Systems (IBS). They are characterized by, among others:
- simplified watchkeeping,
- standardize HMI,
- enhanced conning display,
- multifunction workstations provide any function at any place,
- consistent data available at each workplace,
- health monitoring of system status and performance,
- data quality and sensor selection management,
- intelligent alert management,
- simple to install and upgrade due to open architecture,
- standardized hardware improves logistics of spares,
- standardized software eases configuration and service,
- increased efficiency,
- improved safety,
- cost savings,
- integration of further ship system data and operation (Raytheon Anschütz GmbH. 2009).

2 INTEGRATED BRIDGE SYSTEMS REVIEW

Later presented, System of Navigational-maneuvering Information Presentation is mainly a part of areas of presentation the vessel's location in space and monitoring the parameters of motion systems, forming the so-called Enhanced Conning Display (ECD). Hence, a further part of this paper will only get down to these types of systems.

Referring to the review of existing IBS, presenting the ship's location in space and its motion parameters, it can be noted that the most of them are used by the navigator in the passage in open waters. Examples:
- ECDIS 2D/3D - system mainly designed for presentation of vessel's location in space based primarily on Global Positioning Systems -

GPS/DGPS, very cumbersome in mooring maneuvers, vector coastline often do not coincide with the actual shape of the berths, simplified the waterline of the vessel (Fig. 1);

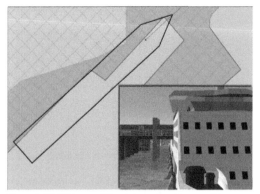

Figure 1. ECDIS – costal line deformation (Own work).

– Radar / ARPA systems – presentation of vessel's location and anti-collision system mainly used in the open sea voyage, not very accurate and inefficient inside the ports (Fig. 2);

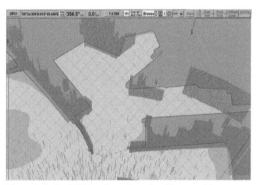

Figure 2. Radar/ARPA echoes distortion (Own work).

– Planning Station / DP systems - expensive positioning systems and presenting of unit's location, which are applied only to the specialist vessels and ferries. Applying them to common units of the merchant fleet is a highly costly and inefficient (Fig. 3).
– Dedicated positioning and presenting of ship's location systems - systems created against order, dedicated for a specific restricted areas or port infrastructure (LNG, Ferries, Narrow Channels). Systems based on highly accurate positioning systems such as Real Time Kinematic (RTK), ladars, ultrasound, which allows navigator to maneuver the ship safely. They are mostly two-dimensional presentation of information systems, with a very simplified model of the ship (Fig. 4).

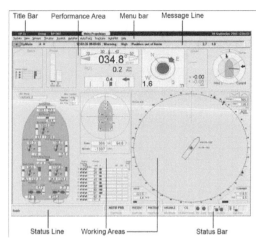

Figure 3. Dynamic Positioning Interface (Kongsberg AS. 2005).

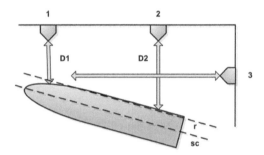

Figure 4. Dedicated LNG positioning system (Gucma, S. & Gucma, M., 2010).

Invariably, for centuries the simplest and most efficient way to assess the vessel's location in space in the restricted areas, in ports is a visual observation.

3 SYSTEM ASSUMPTIONS

No system being supplied with modern navigation bridge is able to completely replace the human factor. Thus, it is necessary to provide the navigator with navigation and maneuvering information as much as possible, and then submit them in an ergonomic form (HMI).

As mentioned utility of systems which presents the position of the vessel against berths in the last phase of passage and during berthing maneuvers is very limited because of its systematic and unsystematic errors.

The only reliable source of information for decision-making is a visual observation. However, it has two basic limitations:
1 is strongly dependent on the currently prevailing hydro-meteorological conditions, particular on the degree of visibility,

2 the navigator has the ability to simultaneously observe only one side of the ship.

For long ship with superstructure in the stern, the navigator on the bridge is not able to assess the distance between the bow section and the obstacle at the height of the waterline. Conversely, the superstructure in the bow - no information about the location of the stern. To perform a safe approach and berthing maneuvers it is necessary to provide information about the navigator position outside of the bridge. Generally it is a crew member equipped with radio communication device or installed on the side of the quay/vessel CCTV camera. Both methods are very limited in heavy fog.

Given the above, decided to create a selective system of navigational-maneuvering information presentation based on mathematical (graphical) models of dedicated vessels and areas. The system allows multi-level assessment of the vessel's location relative to obstacles, regardless of weather conditions.

Main task of system is to faithful reproduction of actual navigational situation with use of mathematical (graphical) models of the vessel and area. Created models were implemented into the virtual environment, where based on standardized data from the positioning systems they are located in the space in three degrees of freedom (in the future the destination is six degrees of freedom).

3.1 Mathematics (graphical) ship's model

Mathematics (graphic) ship model is built in three-dimensional environment based on the technical documentation provided by the unit owner. Virtual hull in both its parts above and below the waterline is a faithful reproduction of a real vessel as to the scale, location and shape (Fig. 5).

Figure 5. Graphical ship's model (Own work).

The degree of detail of the model depends on the complexity of the original hull's shape. There is pos-

sibility to make a virtualization of elements constantly attached to the hull well as moving parts.

3.2 Mathematics (graphical) area's model

Mathematics (graphical) model of areas is also built in three-dimensional environment based on the technical, spatial plans, digital maps (Fig. 6) In the absence of sufficiently accurate plans, it is possible to digitize the virtual model of the basin on the basis of geodetic measurements of such wharves with the use of RTK for instance.

Figure 6. Graphical area's model (Own work).

In addition to the model of the area coastline it is also necessary to make a virtualization of:
- buoys and navigation marks
- hydrotechnical architecture,
- water surface (simplified model).

Thus prepared, the model is positioned in the WGS84 datum of the behavior of the real values of coordinates Lat / Lon or UTM.

4 SYSTEM STRUCTURE

The system consists of two blocks:
1 collecting and recording of input data into system memory
2 reading and data processing

Ad.1. Block of collecting and recording of input data is an independent algorithm that allows to implement to the system variables from independent sources, without requiring changes to the code of second block. This means there is possibility to connect, e.g. various positioning systems, continuous hydro-meteorological data, etc. (Fig. 7).

Manual defining of the fixed input is also available.

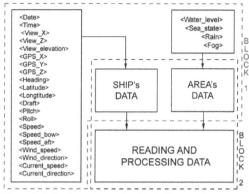

Figure 7. Collecting and recording block (Own work).

Ad.2. Block of reading and data processing is responsible for the division and assignment of the relevant variables to the mathematical (graphical) models of the vessel and area (Fig. 8).

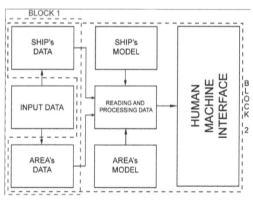

Figure 8. Reading and data processing block (Own work).

In the first version of the software there are available:
1 Data for the model of the ship:
   – the date and time,
   – local coordinates of the point of view and the viewing angle from the bridge,
   – local coordinates of the positioning system's antenna location (it is possible to define more points of reference for the ship's hull),
   – heading,
   – global position of the antenna (vessel's position),
   – draught,
   – the value of trim / pitch (automatic only available with an additional gyro),
   – the value of roll (automatic only available with an additional gyro),
   – the longitudinal and transverse speed,
   – the strength and direction of current / wind.
2 Data for the model of the basin:
   – current water level,

   – sea State (simplified model),
   – intensity of rainfall (automatic only with additional sensors),
   – fog level (automatic only with additional sensors).

Models with the actual data associated, create a virtual interface that reflects the actual navigation-maneuvering situation.

5 SYSTEM INTERFACE

The concept of an interface for the involves the implementation of the following features (Fig. 9):
1 The main screen - a view from the bridge at the centerline of the ship;
2 The navigation bar - the presentation of weather and maneuvering information, with the option of transfer to any location on the screen;
3 System Tray;
4 New camera button - a function that allows simultaneous viewing up to five places in the vicinity of the ship,
5 Preview of added cameras – by clicking on the thumbnails for the camera larger screen is obtained.

Figure 9. Proposal of system interface (Own work).

One of the system's novelty is function, that allows to place the navigator at any point in space around the vessel (including outside the vessel) to 5 virtual cameras simultaneously. With this option it is achieved a full picture of the current navigational situation regardless of weather conditions (Fig. 10)

Each newly-added camera has a thumbnail preview at the bottom of the screen. At any time you can zoom in on a miniature picture, placing it on the screen as the larger windows (Windows® style). In this mode, there is possibility to make moving, rotating, zooming each camera individually. Site selection is made on the two-dimensional map.

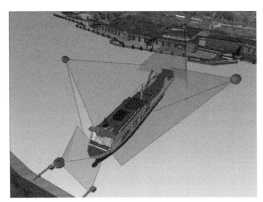

Figure 10. Additional cameras location – turning maneuver (Own work).

The interface system is designed to faithfully reflect the actual navigation and maneuvering as much possible in the virtual environment and, through its innovative features to enhance the safety of the ship maneuvers in a restricted area.

Verification of the graphic interface will undergo a expert tests, in order to improve ergonomics of reading and interpretation of shown information.,

6 CONCLUSION

The proposed decision support system will be developed to improve the safety and to optimize the maneuvering in restricted areas. It was assumed that non-autonomous methods will be used to its verification with a fullmission bridge simulator, as well as selected aspects will be examined in reality (with m/s Navigator XXI).

In the future, the following tests, inter alia, are expected to perform in configuration with and without the proposed system:
- safety maneuver in different weather conditions - safety lines,
- energy hull contact with the fender,
- optimize the number of given orders by a navigator,
- others.

Presented system has a development structure. This means there is possibility to implement the modern features such as: anti-collision system, planning and monitoring of virtual routes, automatic measurement of the CPA to the berth, the prediction power of contact with the fender, etc.

REFERENCES

Gucma, S. & Gucma, M. 2010. Specialized Navigational Systems Used in LNG Terminals. *Scientific Journals Maritime University of Szczecin* 20(92): 45–51. Szczecin.
IMO Resolution MSC.191(79). 2004. Performance standards for the presentation of navigation-related information on shipborne navigational displays.
IMO Resolution MSC.252(83). 2007. Revised performance standards for Integrated Navigation Systems (INS).
IMO MSC/Circ.982. 2000. Guidelines on ergonomic criteria for bridge equipment and layout.
IMO SN/Cir. 243. 2004. Guidelines for the presentation of navigation-related symbols, terms and abbreviations.
IMO SN.1/Circ.274. 2008. Guidelines for the application of the modular concept to performance standards.
Kongsberg AS. 2005. Polaris User Manual. 2005.
Raytheon Anschütz GmbH. 2009. Synapsis Intelligent Bridge Control. http://www.raytheon-anschuetz.com.

12. Security Modeling Technique: Visualizing Information of Security Plans

D. Ley & E. Dalinger
Fraunhofer Institute for Communication, Information Processing and Ergonomics FKIE
Wachtberg, Germany

ABSTRACT: Since the terrorist attacks of 11th September 2001 efforts are made to enhance the security standards in maritime shipping. The joint research project VESPER (improving the security of passengers on ferries) funded by the German Federal Ministry of Education and Research (BMBF) addresses among others the investigation of sea- and landside measures and processes. One of the results of the ongoing analysis phase is that the current representation of information in security plans can hardly be used for a complete real-time implementation of relevant measures in critical situations. As a result, information and design requirements to develop a Security Modeling Technique (SMT) were identified using the Applied Cognitive Work Analysis (ACWA). SMT intends to support decision making of security officers during the implementation of security levels. This contribution describes the phases of development and the resulting design concept.

1 INTRODUCTION

Since the terrorist attacks of 11th September 2001 security has become increasingly important on a world-wide scale and the possibility of terrorist attacks against seagoing ships came to the fore as well. Consequences of such events are immense. A sudden loss of a large number of human lives, destruction of material assets, environmental damages, significant disruptions of transport streams and a possible loss of confidence in maritime infrastructures can be mentioned in this context.

In order to take preventive actions against terroristic attacks, the ISPS Code (International Ship and Port Facility Security Code) was introduced in 2004 (Regulation (EC) No 725/2004) due to the efforts of the International Maritime Organization (IMO). However, the prescribed security management needs further adaptations and optimization. Against this background, the collaborative project VESPER (improving the security of passengers on ferries) was introduced, which is funded by the German Federal Ministry of Education and Research (BMBF). The project addresses the problem of terrorist threats in the maritime domain. The focus is on ferries and passenger ships operating on international routes, especially regarding roll-on roll-off processes, i.e. cargo stevedored via lorries and trains.

The purpose of the collaborative project VESPER is to systematically review the current security standard and to improve hazard prevention measures for ferries. The focus is on security during the access to the ships as well as on the shipboard and seaward measures. To guarantee security standards, there are several originated positions required by the ISPS-Code, including a Ship Security Officer (SSO) and a Port Facility Security Officer (PFSO), which are responsible for identifying threats to the ship or port security, recognizing their significance, and responding to them. Among other intentions the emphasis of VESPER is on optimization of handling processes, (especially a support for the implementation of measures for different security levels). This includes the introduction of aids for decision making in a crisis in order to minimize risks.

Within the framework of VESPER, a new modeling technique to support SSOs and PFSOs was developed, which is described in this contribution.

First, the problem of handling changes of security levels is described (chapter 2). Subsequently, a method for the design of complex systems to support effective decision making, the Applied Cognitive Work Analysis (ACWA), is introduced (chapter 3). The application of ACWA to the described work domain and the consequential outcome, the Security Modeling Technique (SMT), which supports security officers in changing security level, is described (chapter 4). Finally, results are summarized and an outlook to future work is given (chapter 5).

2 PROBLEM FORMULATION

For the purpose of analyzing and optimizing security relevant processes valid process and communication models must be constructed. So, plenty data acquisition methods have been conducted, as observations, exercise participations and interviews and discussions with experts like security officers, designated authorities, port operators, water police officers and finally document analyses like examinations of ship and port facility security plans.

In security plans measures and other information are defined for three different security levels, whereby level one is the level conducted by default. In case of a security relevant event, measures must be preventively intensified or added to a higher security level. Security officers are responsible for a proper and prompt initiation of such security level changes. The analysis of collected data showed that the current presentation of information in security plans relevant for a security level change was not suitable for a prompt and complete implementation of relevant measures in security-critical situations. Measures and information for the three defined security levels are mostly listed in wide and confusing tables and continuous texts in several text passages of a security plan.

Such a presentation of information does not permit to operate efficiently. This weak point has been confirmed by experts in subsequent discussions and by the data analysis of the accompanied exercise.

Hence, a demand for an optimized information representation and therefore information management has been identified. This should support security officers in making decisions. For this purpose a concept of a new modeling technique has been developed based on the Applied Cognitive Work Analysis (ACWA), which is introduced in the following part.

3 METHODICAL APPROACH

In order to develop a modeling technique for a support of decision making processes of security officers, first, an analysis of the work domain must be performed. The traditional task analysis methods focus on what operators do and what tasks must be fulfilled and provide descriptions of task sequences (Annett, 2004; Kirwan and Ainsworth, 1992). To account for factors like unanticipated events, dynamic changes of the situation, and real-time reactions to these changes, methods are required, which examine human cognitive processes. Cognitive Systems Engineering (CSE) is a design framework which focuses on analysis of cognitive demands in order to identify cognitive processes of operators (Crandall et al., 2006; Rasmussen et al., 1994). Methods of CSE help to understand, how experts make decisions and why

they make certain decisions, what cues they need, what knowledge and strategies they use. The Applied Cognitive Work Analysis (ACWA) is a CSE approach for the analysis, design and evaluation of complex systems and interfaces. In this paper we discuss the application of ACWA for designing a modeling technique.

ACWA is a methodology for the design of a user interface for effective decision support. The process begins with the identification of the decisions that operators must make and ends with the identification of visualization and decision-aiding concepts. Thus, this methodology can be used to develop a technique to support decision maker, which is based on effective information visualization.

ACWA comprises the following process steps (Elm at al., 2003, see figure 1):

– Development of a Functional Abstraction Network (FAN) – a model to represent the functional relationships between the work domain elements
– Identification of cognitive demands which arise in the domain and need support – Cognitive Work Requirements (CWR) or decision requirements
– Identification of the Information/Relationship Requirements (IRR) for effective decision-making
– Definition of a relationship between the decision requirements and visualization concepts (how the information needs to be represented) – Representation Design Requirements (RDR)
– Implementation of representation requirements into a powerful visualization of the domain context – Presentation Design Concepts (PDC)

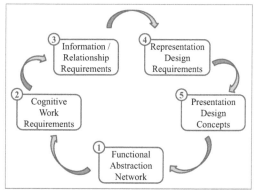

Figure 1. Iterative steps of the Applied Cognitive Work Analysis (ACWA).

The ACWA begins with a FAN which is a function-based goal-means decomposition of the domain, based on Rasmussen's representation formalism for a work domain – an abstraction hierarchy (Rasmussen, 1985), which describes human information processing. With ascending in the hierarchy the understanding for goals to achieve rises. Moving to deeper levels reveals a better understanding for the system's functions with a view to achievement of these goals. The FAN is a multi-level representation of the work domain. Each node in the network represents a goal, links represent support. Each goal has a process providing a description how to achieve this goal. Processes define supporting functions for achieving the goals in the hierarchal level above.

The FAN provides the basis for the definition of CWR or decision requirements. The CWR help to gain understanding of the goals in the work domain and enhance the decision-centred perspective. The decision requirements are to be defined for each goal node in the FAN. This ensures an understanding what decisions are to be made to achieve the goals.

Next step is to identify required information for each decision. Factors, which are essential for decision making, are identified with the CWR and therefore the context for information requirements is provided. Decision making is based upon the interpretation of information. Incorrect or incomplete information leads to wrong decisions.

Hence, the way of information presentation is very important. Appropriate information visualization can improve information processing and thus the process of decision making. The next step is to develop the decision-aiding concepts on the basis of the information requirements taking into account human perception and cognition. Display concepts which support the cognitive tasks through an appropriate visualization should be developed. At this step several different design concepts may be generated. These design concepts are still requirements and not an implementation.

The developed visualization concepts provide hypotheses about effective decision support. The next step is the development of a prototype to evaluate the effectiveness of the new system. The prototype can help to identify additional decision and information requirements for decision support which have been missing in the first steps. Thus, the ACWA approach is an iterative process (see figure 1), which leads in several steps from the analysis of the demands of the work domain to the identification of effective decision-aiding visualizations.

4 APPLYING ACWA

Subsequent it is described how the ACWA approach can be applied to determine design requirements for a security modeling technique. To gain understanding of the domain of maritime security diverse knowledge elicitation techniques have been used. Relevant documents, such as the ISPS Code, ship security plans and safety management handbooks for ships have been reviewed, interviews with masters and security officers on board of ferries have been conducted.

4.1 *Functional Abstraction Network (FAN)*

Based on collected information a FAN, which is the first step of ACWA (Figure 1), has been generated. The rectangles (Figure 2) represent goals, which are organized hierarchically, links represent supporting connections from lower-level goals to higher-level goals. The achievement of goals is described through processes (not represented in the figure).

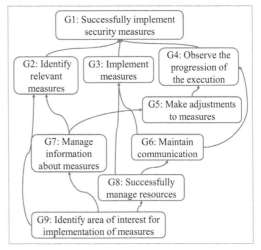

Figure 2. Functional Abstraction Network (FAN)

The overall goal is to successfully implement security measures from a ship security plan. For example, this goal can be described as a process as follows: Initial situation is a declared change of a security level. This induces the responsible security officer to find appropriate measures for a level change. Next, these measures have to be initiated accordingly. In the further process step the accomplishment of measures has to be controlled until all relevant measures are implemented. The overall goal of implementing security measures is supported by subordinated goals (with corresponding process descriptions). Their goals are described below.

In order to manage successfully the change of a security level one needs to identify the appropriate measures. In addition, measures have to be implemented and the progression of the execution of measures has to be observed. The goal of the observation process is to determine, whether the implementation of measures is working as intended, and whether the situation has changed or not. The observation should proceed continuously to assure the notification of failures. If something goes wrong, the implemented measures fail or changes in the situation are identified, then it will be possible to correct it in sufficient time. After this, the new measures can be defined and implemented.

Table 1. Exemplary requirements for goals 7 to 9

Goal	CWR	IRR	RDR
G7	- Choose necessary information on measures - Search for additional information concerning measures - Recognize passenger procedures - …	- Indicate security levels - Indicate means/tools - Indicate allocation of means/tools to measures in different security levels - Indicate comments - Indicate belonging of comments - …	Represent security level I in green colour Represent security level II in yellow color Represent security level II in red color …
G8	- Select operator necessary to fulfill tasks - Allocate operator to area - Select technical resources necessary to fulfill tasks - Allocate technical resources necessary to fulfill tasks	- Assignment of operators to measures - Necessary means/tools - Assignment of technical resources to measures - …	- Integrate operator labels in measure shapes - Separated representation of means/tools - Connection of technical resources with appropriate measures of a security level - …
G9	- Choose relevant areas for conducting measures - …	- Information about the kind of areas (ships, port facilities, sub-areas, restricted areas, permitted areas) and their locations - Caption of areas	- Represent ships with abstracted contour - Represent port facilities with abstracted contour

Information about measures has to be managed, which is a supporting goal for the measure adjustments as well as the identification of relevant measures. Communication plays an important role in gaining and forwarding information. One needs to provide the information of the incident to responsible personnel on board and receive the responses.

Maintenance of communication supports the implementation and observation of execution of measures. On the other hand communication maintenance and implementation of measures are achieved through a successful management of resources. Before the initiation of countermeasures it is essential to determine whether the resources (including personnel and equipment) necessary to fulfill the measures are available or adequate. Finally, the identification of the area of interest is necessary for the implementation of measures, supporting the identification of relevant measures and the management of information as well as resources.

4.2 Cognitive Work Requirements (CWR)

The developed FAN is the basis for the definition of Cognitive Work Requirements (CWR) respectively decision requirements. Cognitive Work Requirements refer to the goals in the FAN. They enable comprehension of what decisions are to be made to reach the defined goals. By this means, CWR help to develop a decision-centered perspective. In the second column of Table 1 CWR are listed for the goals G7, G8 and G9 of the FAN (Figure 2). For example,

for the goal "Successfully manage resources" (G8) an operator and technical resources have to be selected. In addition the operator and the technical resources have to be allocated to the relevant area.

4.3 Information/Relationship Requirements (IRR)

Decision making is based on interpretation of information. Incorrect or incomplete information leads to incorrect decisions. Thus, the next step of ACWA is to identify information necessary to come to decisions. In the third column of Table 1 information/relationship requirements are listed with reference to the CWR. For example, the goal "Choose relevant areas for conducting measures" (G9) requires to provide information about the kind of areas (e.g. ships and ports), as well as caption of areas.

4.4 Representing Design Requirements (RDR)

Next, representing design requirements (RDR) based on the IRR are defined, constituting first design ideas for an effective decision making support. The kind of information representation is of great significance since an adequate visualization of information can improve human information processing and with that the process of decision making. In the right column of table 1 RDR are listed. For example, the indication of security levels is implemented in terms of the colours green, yellow and red in order to fulfil G7: "Manage information about measures".

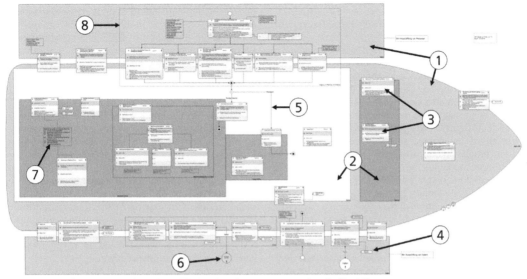

Figure 3. Exemplary SMT model of a ferry ship (numbered components are described in 4.5)

4.5 *Presentation Design Concepts (PDC)*

The next step of the ACWA method involves a draft of a presentation design concept and its prototypical implementation. Since ACWA is an iterative approach a prototype enables the developer to define additional requirements for the next iteration not identified in the previous one. A prototype is currently under development. Below, the presentation design concepts are introduced.

Basically, the SMT consists of five modeling categories integrated in one complex model: Modeling of areas, security levels, measures, processes and communication. In order to simplify the application of SMT, the quantity of components has been minimized. Figure 3 shows a SMT model of a ferry ship, representing information of a ship security plan. Such a model, including interfaces to adjacent port areas, allows the responsible ship security officer to get an overview of all relevant measures and adjoining information for a change of a security level. In the following we will explain the five modeling categories, mentioned above (numbers correspond to figure 3).

Modeling of areas: Fundamental to a SMT model is a modeling of areas corresponding to the real spatial conditions. Basically, there are shapes representing ships and port facilities (1) as well as shapes representing their corresponding sub-areas, namely restricted (grey) and permitted (white) areas (2). By this means, the categorization of areas concerning access authorizations is immediately evident as well as the search for location-dependent information is simplified. Particularly, the interface between ship and port which has been neglected so far, can now

be taken into account by the use of the concept of area modeling.

Modeling of security levels: Each measure is allocated to one of three defined security levels. To find measures and further information for a specific security level, an intuitive and differentiating illustration is needed. Therefore, information related to security level one is coded green, to security level two yellow and to security level three red. Additionally, security levels are identifiable through Roman numerals. Apart from measures, also contact points and resources are modeled depending security levels. See Figure 4 for an activity shape representing the security levels.

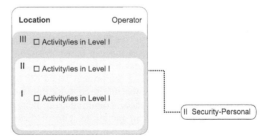

Figure 4: Activity shape

Modeling of measures: Measures are modeled in activity shapes for a certain area or the access of an area (3). An activity shape is structured in the following way (see Figure 4). At the top left is a label for the area for which the measures have to be applied. At the top right a responsible operator can be named. Apart from that the shape is divided into three colored sections listing the measures for the three security levels. Thereby, measures of higher levels substitute those of lower levels, visualized through the integration of lower level fields into higher ones. Checkboxes in front of each measure allow the marking of implemented measures.

Specific tools, equipment or additional personnel are partially necessary for a successful implementation of measures. To visualize the demand for such resources, a resource shape (4) is added to the corresponding measure (see resource shape for security level II in figure 4). By this means a resource management is maintained so that available resources can be found and allocated situation-dependently.

Process modeling: In most cases access controls for different passenger categories must be conducted in a predefined manner. In case of an occurrence of such structured procedures, concerned activity shapes can be connected by control flows and thus represent processes (5). By modeling processes the security officer deploying the model is able to quickly identify measures in chronological order and may coordinate involved measures according to the particular situation. Beginning and ending of a process are represented through distinct circles.

Communication modeling: Communication between different points is a condition for a successful implementation of measures and coordination of resources. Hence, given communication and information paths between appropriate contact points must be presented in SMT models. This is realized through contact point shapes and corresponding information flows (6).

Furthermore, there are additional components like a shape for the insertion of explanations (7) and a shape for a clustering of content-related components for clarity improvement (8).

5 SUMMARY AND FUTURE WORK

The Applied Cognitive Work Analysis (ACWA; Elm et al., 2003) is a Cognitive Systems Engineering (CSE) method which closes the gap between cognitive analysis and design existing in other methods. ACWA has been applied in the project VESPER to get a functional model of the maritime work domain of security officers. Hence, cognitive and information demands have been identified. Out of these demands visualization and design requirements have been derived. These previous steps provided the basis for the development of a presentation design concept of the security modeling technique. This technique, called 'SMT', enables security officers to create models of ships and port facilities. These models, used as a computer-based tool or as a large-format poster, support them in making decisions

during the implementation of measures and the management of resources in the context of a security level change. Emphasis of SMT is a suitable representation of security plan information. It illustrates spatial conditions, communication, processes and area-specific measures in an integrated manner and also distinguishes the three defined security levels.

To ensure a user-friendly development of SMT models by ship, company and port facility security officers but also officers of designated authorities, an SMT editor is currently under development within the iterative development process of ACWA. The editor will also contain control functions, e.g. to guarantee the completeness of modeling measures.

The concept of SMT has been developed in close collaboration with experts in the field of maritime security (e.g. masters, security officers, ship companies, officers of designated authorities). Also the SMT editor will be evaluated and improved in intensive cooperation with these experts.

Moreover, the application of SMT and the editor shall not be limited to ferry shipping. Therefore, the concept has to be tailored to the entire international shipping. Feedback of involved experts of the maritime security domain, including representatives of German designated authorities and delegates of the European Commission, shows a concordant endorsement of the use of SMT models.

REFERENCES

Annett, J. 2004. Hierarchical Task Analysis. In D. Diaper & N. Stanton (eds.), The Handbook of Task Analysis for Human-Computer Interaction: 67-82. Mahwah, New Jersey: Erlbaum.

Crandall, B., Klein, G. & Hoffman, R.R. 2006. Working Minds: A Practitioner's Guide to Cognitive Task Analysis. Cambridge, MA: MIT Press.

Dalinger, E. & Motz, F. 2010. Designing a decision support system for maritime security incident response. In O. Turan, J. Boss, J. Stark, J.L. Colwell (eds). Proceedings of International Conference on Human Performance at Sea.

Elm, W., Potter, S., Gualtieri, J., Roth, E. & Easter, J. 2003. Applied cognitive work analysis: a pragmatic methodology for designing revolutionary cognitive affordances. In E. Hollnagel (ed.), Handbook for Cognitive Task Design, London: Lawrence Erlbaum Associates.

Kirwan, B. & Ainsworth, L. K. 1992. A Guide to Task Analysis. London: Taylor & Francis.

Rasmussen, J., Pejtersen, A.M. & Goodstein, L.P. 1994: Cognitive Systems Engineering. New York: Wiley.

Rasmussen, J. 1985. The Role of Hierarchical Knowledge Representation in Decisionmaking and System Management. IEEE Transactions on Systems, Man, and Cybernetics. SMC-15 (2): 234-243.

Regulation (EC) No 725/2004 of the European Parliament and of the Council of 31 March 2004 on enhancing ship and port facility security'. Official Journal L 129 of 29.04.2004.

Data Transmission and Communication Systems

Data Transmission and Communication Systems

International Recent Issues about ECDIS, e-Navigation and Safety at Sea – Marine Navigation and Safety of Sea Transportation – Weintrit (ed.)

13. Maritime Communication, Navigation and Surveillance (CNS)

S. D. Ilcev

Durban University of Technology (DUT), Durban, South Africa

ABSTRACT: This paper introduces development and implementation of Maritime Satellite Communications, Navigation and Surveillance (CNS) of GPS or GLONASS for enhancement of safety and emergency systems including security and control of vessels, logistic and freight at sea, on inland waters and the security of crew and passengers on board ships, cruisers, boats, rigs and hovercrafts. These improvements include many applications for the better management and operation of vessels and they are needed more than ever because of world merchant fleet expansion. Just the top 20 world ships registers have more than 40,000 units under their national flags. Above all, the biggest problem today is that merchant ships and their crews are targets of the types of crime traditionally associated with the maritime industries, such as piracy, robbery and recently, a target for terrorist attacks. Thus, International Maritime Organization (IMO) and flag states will have a vital role in developing International Ship and Port Security (ISPS). The best way to implement ISPS is to design an Approaching and Port Control System (APCS) by special code augmentation satellite CNS for all ships including tracking and monitoring of all vehicle circulation in and out of the seaport area. The establishment of Maritime CNS is discussed as a part of Global Satellite Augmentation Systems (GSAS) of the US GPS and Russian GLONASS for integration of the existing Regional Satellite Augmentation Systems (RSAS) such as the US WAAS, European EGNOS and Japanese MSAS, and for development new RSAS such as the Russian SDMC, Chinese SNAS, Indian GAGAN and African ASAS. This research has also to include RSAS for Australia and South America, to meet all requirements for GSAS and to complement the services already provided by Differential GPS (DGPS) for Maritime application of the US Coast Guard by development Local Satellite Augmentation System (LSAS) in seaports areas.

1 INTRODUCTION

The current infrastructures of the Global Navigation Satellite System (GNSS) applications are represented by old fundamental solutions for Position, Velocity and Time (PVT) of the satellite navigation and determination systems such as the US GPS and Russian (former-USSR) GLONASS military requirements, respectively. The GPS and GLONASS are first generation of GNSS-1 infrastructures giving positions to about 30 metres, using simple GPS/GLONASS receivers (Rx) onboard ships or aircraft, and they therefore suffer from certain weaknesses, which make them impossible to be used as the sole means of navigation for ships, particularly for land (road and rail) and aviation applications. In this sense, technically GPS or GLONASS systems used autonomously are incapable of meeting civil maritime, land and especially aeronautical mobile very high requirements for integrity, position availability and determination precision in particular for Traffic Control and Management (TCM) and are insufficient for certain very critical navigation and flight stages. [01, 02].

Because these two systems are developed to provide navigation particulars of position and speed on the ship's bridges or in the airplane cockpits, only captains of the ships or airplanes know very well their position and speed, but people in Traffic Control Centers (TCC) cannot get in all circumstances their navigation or flight data without service of new CNS facilities. Besides of accuracy of GPS or GLONASS, without new CNS is not possible to provide full TCM in every critical or unusual situation. Also these two GNSS systems are initially developed for military utilization only, and now are also serving for all transport civilian applications worldwide, so many countries and international organizations would never be dependent on or even entrust people's safety to GNSS systems controlled by one or two countries. However, augmented GNSS-1 solutions of GSAS network were recently developed to improve the mentioned deficiencies of current military systems and to meet the present transportation civilian requirements for high-operating Integrity, Continuity, Accuracy and Availability (ICAA). These new developed and operational CNS solutions are the US Wide Area Augmentation System (WAAS), the European Geostationary Navigation Overlay System (EGNOS) and Japanese MTSAT Satellite-based Augmentation System (MSAS), and there are able to provide CNS data from mobiles to the TCC via Geostationary Earth Orbit (GEO) satellite constellation.

These three RSAS are integration segments of the GSAS network and parts of the interoperable GNSS-1 architecture of GPS and GLONASS and new GNSS-2 of the European Galileo and Chinese Compass, including Inmarsat CNSO (Civil Navigation Satellite Overlay) and new projects of RSAS infrastructures. The additional four RSAS of GNSS-1 networks in development phase are the Russian System of Differential Correction and Monitoring (SDCM), the Chinese Satellite Navigation Augmentation System (SNAS), Indian GPS/GLONASS and GEOS Augmented Navigation (GAGAN) and African Satellite Augmentation System (ASAS). Only remain something to be done in South America and Australia for establishment of the GSAS infrastructure globally, illustrated in Figure 1.

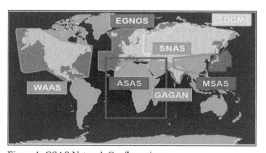

Figure 1. GSAS Network Configuration
Courtesy of Book: "Global Aeronautical CNS" by Ilcev [01]

The RSAS solutions are based on the GNSS-1 signals for augmentation, which evolution is known as the GSAS network and which service provides an overlay function and supplementary services. The future ASAS Space Segment will be consisted by existing GEO birds, such as Inmarsat-4 and Artemis or it will implement own satellite constellation, to transmit overlay signals almost identical to those of GPS and GLONASS and provide CNS service. The South African firm IS Marine Radio, as designer of the Project will have overall responsibility for the design and development of the ASAS network with all governments in the region.

1.1 GNSS Applications

The RSAS infrastructures are available globally to enhance current standalone GPS and GLONASS

system PVT performances for maritime, land (road and rail) and aeronautical transport applications. User devices can be configured to make use of internal sensors for added robustness in the presence of jamming, or to aid in vehicle navigation when the satellite signals are blocked in the "urban canyons" of tall city buildings or mountainous environment. In the similar sense, some special transport solutions, such as maritime and especially aeronautical, require far more CNS accuracy and reliability than it can be provided by current military GPS and GLONASS space infrastructures [01, 03].

Moreover, positioning accuracy can be improved by removing the correlated errors between two or more satellites GPS and/or GLONASS Rx terminals performing range measurements to the same satellites. This type of Rx is in fact Reference Receiver (RR) surveyed in, because its geographical location is precisely well known. In such a manner, one method of achieving common error removal is to take the difference between the RR terminals surveyed position and its electronically derived position at a discrete time point. These positions differences represent the error at the measurement time and are denoted as the differential correction, which information may be broadcast via GEO data link to the user receiving equipment. In this case the user GPS or GLONASS augmented Rx can remove the error from its received data.

Alternatively, in non-real-time technique GNSS solutions, the differential corrections can be stored along with the user's positional data and will be applied after the data collection period, which is typically used in surveying applications [04].

If the RR or Ground Monitoring Station (GMS) of the mobile users, the mode is usually referred to as local area differential, similar to the US DGPS for Maritime applications. In this way, as the distance increases between the users and the GMS, some ranging errors become decorrelated. This problem can be overcome by installing a network consisting a number of GMS reference sites throughout a large geographic area, such as a region or continent and broadcasting the Differential Corrections (DC) via GEO satellites. In such a way, the new projected ASAS network has to cover entire African Continent and the Middle East region.

Therefore, all GMS sites connected by Terrestrial Telecommunication Networks (TTN) relay collected data to one or more Ground Control Stations (GCS), where DC is performed and satellite signal integrity is checked. Then, the GCS sends the corrections and integrity data to a major Ground Earth Station (GES) for uplink to the GEO satellite. This differential technique is referred to as the wide area differential system, which is implemented by GNSS system known as Wide Augmentation Area (WAA), while another system known, as Local Augmentation Area

(LAA) is an implementation of a local area differential [05].

The LAA solution is an implementation for seaports and airport including for approaching utilizations. The WAA is an implementation of a wide area differential system for wide area CNS maritime, land and aeronautical applications, such as Inmarsat CNSO and the newly developed Satellite Augmentation WAAS in the USA, the European EGNOS and Japanese MSAS [03].

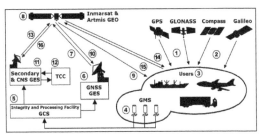

Figure 2. ASAS Network Configuration
Courtesy of Book: "Understanding GPS - Principles and Applications" by E.D. Kaplan [03]

These three operational systems are part of the worldwide GSAS network and integration segments of the future interoperable GNSS-1 architecture of GPS and GLONASS and GNSS-2 of Galileo and Compass, including CNSO as a part of GNSS offering this service via Inmarsat-3/4 and Artemis spacecraft. The author of this paper for the first time is using more adequate nomenclature GSAS than Satellite-based Augmentation System (SBAS) of ICAO, which has to be adopted as the more common designation in the field of CNS [06].

As discussed earlier, the current three RSAS networks in development phase are the Russian SDCM, Chinese SNAS and Indian GAGAN, while African Continent and Middle East have to start at the beginning of 2011 with development ASAS project. In this sense, development of forthcoming RSAS projects in Australia and South America will complete Augmented CNS system worldwide, known as an GSAS Network [04].

Three operational RSAS together with Inmarsat CNSO are interoperable, compatible and each constituted of a network of GPS or GLONASS observation stations and own and/or leased GEO communication satellites. Namely, the Inmarsat CNSO system offers on leasing GNSS payload to the European system EGNOS, which will provide precision to within about 5 metres and is operational from 2009. In fact, it also constitutes the first steps towards forthcoming Galileo, the future European system for civilian global navigation by satellite. The EGNOS system uses leased Inmarsat AOR-E and IOR satellites and ESA ARTEMIS satellite. Thus, the US-based WAAS is using Inmarsat satellites and

Japanese MSAS is using its own multipurpose MTSAT spacecraft, both are operational from 2007 and 2008, respectively. Although the global positioning accuracy system associated with the overlay is a function of numerous technical factors, including the ground network architecture, the expected accuracy for the US Federal Aviation Administration (FAA) WAAS will be in the order of 7.6 m (2 drms, 95%) in the horizontal plane and 7.6 m (95%) in the vertical plane [04, 05].

1.2 RSAS System Configuration

The RSAS network are designed and implemented as the primary means of satellite CNS for maritime course operations such as ocean crossings, navigation at open and close seas, coastal navigation, channels and passages, approachings to anchorages and ports, and inside of ports, and for land (road and railways) solutions. In this sense, it will also serve for aviation routes in corridors over continents and oceans, control of airports approachings and managing all aircraft and vehicles movements on airports surface [03]:

It was intended to provide the following services:

1. The transmission of integrity and health information on each GPS or GLONASS satellite in real time to ensure all users do not use faulty satellites for navigation, known as the GNSS Integrity Channel (GIC).

2. The continuous transmission of ranging signals in addition to the GIC service, to supplement GPS, thereby increasing GPS/GLONASS signal availability. Increased signal availability also translates into an increase in Receiver Autonomous Integrity Monitoring (RAIM) availability, which is known as Ranging GIC (RGIC).

3. The transmission of GPS or GLONASS wide area differential corrections has, in addition to the GIC and RGIC services, to increase the accuracy of civil GPS and GLONASS signals. Namely, this feature has been called the Wide Area Differential GNSS (WADGNSS).

The combination of the Inmarsat overlay services and Artemis spacecraft will be referred to as the ASAS network illustrated in Figure 2. As observed previous figure, all mobile users (3) receive navigation signals (1) from GNSS-1 of GPS or GLONASS satellites. In the near future can be used GNSS-2 signals of Galileo and Compass satellites (2). These signals are also received by all reference GMS terminals of integrity monitoring networks (4) operated by governmental agencies in all countries within Africa and Middle East.

The monitored data are sent to a regional Integrity and Processing Facility of GCS (5), where the data is processed to form the integrity and WADGNSS correction messages, which are then forwarded to the Primary GNSS GES (6). At the GES, the navigation signals are precisely synchronized to a reference time and modulated with the GIC message data and WADGNSS corrections. The signals are sent to a satellite on the C-band uplink (7) via GNSS payload located in GEO Inmarsat and Artemis spacecraft (8), the augmented signals are frequency-translated to the mobile user on L1 and new L5-band (9) and to the C-band (10) used for maintaining the navigation signal timing loop. The timing of the signal is done in a very precise manner in order that the signal will appear as though it was generated on board the satellite as a GPS ranging signal. The Secondary GNSS GES can be installed in Communication CNS GES (11), as a hot standby in the event of failure at the Primary GNSS GES. The TCC ground terminals (12) could send request to all particular mobiles for providing CNS information by Voice or Data, including new Voice, Data and Video over IP (VDVoIP) on C-band uplink (13) via Communication payload located in Inmarsat or Artemis spacecraft and on C-band downlink (14) to mobile users (3). The mobile users are able to send augmented CNS data on L-band uplink (15) via the same spacecraft and L-band downlink (16). The TCC sites are processing CNS data received from mobile users by Host and displaying on the surveillance screen their current positions very accurate and in the real time [03]. Therefore, the ASAS will be used as a primary means of navigation during all phases of traveling for all mobile applications [06].

The RSAS space constellation could be formally consisted in the 24 operational GPS and 24 GLONASS satellites and of 2 Inmarsat and 1 Artemis GEO satellites. The GEO satellites downlink the data to the users on the GPS L1 RF with a modulation similar to that used by GPS. Information in the navigational message, when processed by an RSAS Rx, allows the GEO satellites to be used as additional GPS-like satellites, thus increasing the availability of the satellite constellation. At this point, the RSAS signal resembles a GPS signal origination from the Gold Code family of 1023 possible codes (19 signals from PRN 120-138).

2 MARITIME TRANSPORTATION AUGMENTATION SYSTEM (MTAS)

The navigation transponder of GEO payload is a key part of the entire system. Thus, it sends GNSS signals to mobiles in the same way as GPS or GLONASS satellites and improves the ICAA positioning system. Thanks to the large number of mobiles, the GNSS signal is able to incorporate data on GPS spacecraft status and correction factors, greatly improving the reliability and accuracy of the present GPS system, which comes to few tenths of metres. The augmented GPS and GLONASS accuracy will be just a few metres, allowing maritime and land

traffic to be controlled solely by satellite, without ground radar or radio beacons facilities.

To complement the GPS channel, communication channels allow bidirectional transmission between ships and GES. The ship sends its position and navigation data to the Port authorities, TTC and to the relevant ship-owner. This enables ship movements to be managed and to enhance safety at sea and to improve operating efficiency. The satellite will forward flexible and safe routing information to ships, as determined by the shore centre, decreasing fuel consumption, reducing sailing times and enhancing the safety and security systems in all sailing stages. The CNS/MTAS mission is divided into three Maritime CNS systems, such as Communication, Navigation and Surveillance. As usual, the MTAS system consists in space and ground infrastructures [2].

2.1 Space Segment

The space segment for MTAS infrastructure and mission, as a part of GSAS configuration, can be the same new designed GEO and/or leased Inmarsat, Japanese MTSAT, European Artemis of ESA or any existing GEO with enough space for GNSS transponder inside of payload. The spacecraft GNSS payload can provide global and spot beam coverage with determined position on about 36,000 km over the equator.

The MTAS spacecraft also can have an innovative communication purpose payload for Maritime Mobile Satellite Service (MMSS), which will be similar to the Inmarsat system of Mobile Satellite Communications (MSC). The heart of the payload is an IF processor that separates all the incoming channels and forwards them to the appropriate beam in both directions: forward (ground-to-ship) and return (ship-to-ground). In fact, global beam covers 1/3 of the Earth between 75° North and South latitudes. Thus, spot beam coverage usually consists in 6 spot beams over determined regions including heavy traffic areas at sea, to meet the demands of increasing maritime transport operations and for enhanced safety and security [6].

The GNSS signal characteristics are generally based on the ICAO Annex 10 (SARP), IMO and Inmarsat SDM and comply with the Radio Regulations and ITU-R Recommendations. This type of spacecraft has two the following types of satellite links related to the maritime Ship Earth Stations (SES) and Ground Earth Stations (GES):

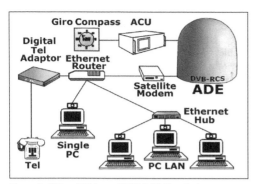

Figure 3. SES or Shipborne DVB-RCS Terminal
Courtesy of Book: "Global Mobile Satellite Communications"
by Ilcev [6]

2.1.1 Forward GES to Satellite Direction

The GES terminals are located throughout the region coverage and their signals are received by L, Ku or a Ka-band ships antenna. Thanks to the very high Radio Frequency (RF) used, the reflector size of the antennas is quite small, 500 mm for Ku-band, 450 mm for Ka-band and double size for L-band. The reflector onboard mobile is movable via focusing tracking motors automatically correcting Azimuth and Elevation angles. The focusing motors are connected to the Gyrocompass onboard ships, so that it can work with the communications satellite payloads in any of the possible vessel positions in four GEO coverages, see Figure 3. The GES uses C-band feeder link and SES uses L-band service link with larger size of antenna than antennas using Ku and Ka-band. The SES standards are using new broadband technique and are capable to provide Broadcast, Multimedia and Internet service for Voice, Data and Video over IP (VDVoIP) and IPTV. Incoming signals are then amplified, converted to IF, filtered and routed within the IF processor where they are then up-converted and transmitted to the SES. Otherwise, the author of this paper proposed this solution in 2000 in his book [6] as Maritime Broadband, seven years before Inmarsat offered and promoted its FleetBroadband.

2.1.2 Return Satellite to GES Direction

The L-band signal received from approaching SES are processed in the same way and retransmitted to GES via Ku and Ka-band GES antennas, although the GES system can also employ Inmarsat C-band transmitter and antenna. The output power of the Ku and Ka-band SES transmitters is just 2W thanks to the high gain satellite antenna. It is also possible to provide station-to-station channels in either the Ku or Ka-band to enable stations working with different spots to communicate with one another. The GNSS channel is also routed to GES on same two bands for calibration purposes [7].

2.2 Ground Segment

The MTAS Ground Segment consists in several GES and Ground Control Terminal (GCT) located in any corresponding positions. Thus, an important feature of these stations is that they have been built to withstand earthquakes, which also required a special antenna design.

2.2.1 Ground Earth Stations (GES)

In order to provide continuous service, even during natural disasters, two GES can be implemented at two different locations separated by about 500 km. The MMSS provided by GES is in charge of all communication functions via satellites. With a 13 m antenna diameter GES transmits and receives signals in the Ku, Ka and C-band. A very high EIRP of 85 dBW and a high G/T ratio of 40 dB/K are achieved in the Ku and Ka-band, respectively and ensure very high availability of the feeder link. The L-band terminal similar to the SES is used for the system testing and monitoring. About 300 circuits are available simultaneously in both: transmit and receive directions. It also includes dedicated equipment for testing the satellite performance after launch and for permanent monitoring of the traffic system. Top-level management software is provided to configure the overall system and check its status.

2.2.2 Ship Earth Stations (SES)

Special part of the MTAS Ground Segment are SES terminals approaching to the entire region including GNSS. It is similar to the Inmarsat standards containing: ADE (Above Deck Equipment) as an antenna and BDE (Below Deck Equipment) as a transceiver with peripheral equipment using L-band. The BDE Voice, Data and Video (VDV) terminals can be used for ship crew and cabin crew including passenger applications. The SES is a ship-mounted radio capable of communications via spacecraft in the MTAS system, providing VDV and Fax two-way service anywhere inside the satellite footprint.

2.2.3 Satellite Control Stations (SCS)

The SCS terminal is usually located in the same building as the GES and utilizes an antenna with the same diameter. This station has to control the satellite throughout its operational life in the Network. Two Radio Frequency (RF) bands can be used: S-band in normal operation and Unified S-band (USB) while the satellite is being transferred to its final orbit, or in the event of an emergency when satellite loses its altitude. Accordingly, in S-band the EIRP is 84 dBW and for security reasons, the EIRP in USB is as high as 104 dBW. An SCS displays the satellite's status and prepares telecommands to the satellite. Furthermore, the satellite position is measured very accurately (within 10 m) using a trilateral ranging system instead of measuring one signal, which is sent to the satellite then returned to the Earth. On the other hand, the Station sends out two additional signals, which are retransmitted by the satellite to two dedicated ranging stations on the ground, which return the same signals to the SCS via satellite. This technique allows the satellite's position to be measured in three dimensions. On the other hand, a dynamic spacecraft simulator is also provided to check telecommands.

2.2.4 *GNSS System*

The GNSS system known as the MTAS for maritime applications consists in a large number of GMS, GCS, GES and few Geostationary Ranging Stations (GRS) to implement a wide triangular observation base for GEO satellite ranging. The GMS terminals are very small autonomous sites housed in a shelter of some adequate building with appropriate antenna system and trained staff. Each GMS computes its location using GPS and MTAS communication signals over the coverage area. Any differences between the calculated and real locations are used by the system to correct the satellite data. Data is sent to the GCS via the public network or satellite links, while the GCS collects all the information from each GMS. Complex software is able to calculate accurately the position and internal times of all GPS and MTAS satellites. The GNSS signal, incorporating the status of the GPS spacecraft and corrections, is calculated and sent to the traffic station known as GES for transmission to MTAS satellites [7].

3 COMPARISON OF THE CURRENT AND NEW MARITIME CNS SYSTEM

Business or corporate shipping and airways companies have used for several decades HF communication for long-range voice and telex communications during intercontinental sailing and flights. Meanwhile, for short distances mobiles have used the well-known VHF onboard ships and VHF/UHF radio on aircraft. In the similar way, data communications are recently also in use, primarily for travel plan and worldwide weather (WX) and navigation (NX) warning reporting. Apart from data service for cabin crew, cabin voice solutions and passenger telephony have also been developed. Thus, all mobiles today are using traditional electronic and instrument navigations systems and for surveillance facilities they are employing radars.

The current communication facilities between ships and Maritime Traffic Control (MTC) are executed by Radio MF/HF voice and telex and by VHF voice system; see Previous Communication Subsystem in Figure 4. The VHF link between ships on one the hand and Coast Radio Station (CRS) and TCC on the other, may have the possibility to be interfered with high mountainous terrain and to provide problems for MTC. The HF link may not be established due to lack of available frequencies, high frequency jamming, bad propagation, intermediation, unstable wave conditions and to very bad weather, heavy rain or thunderstorms.

The current navigation possibilities for recording and processing Radio Direction Information (RDI) and Radio Direction Distance Information (RDDI) between vessels and TCC or MTC centre are performed by ground navigation equipment, such as the shore Radar, Racons (Radar Beacon) and Passive Radar Reflectors, integrated with VHF CRS facilities, shown by Previous Navigation Subsystem in Figure 4. However, this subsystem needs more time for ranging and secure navigation at the deep seas, within the channels and approachings to the anchorages and ports, using few onboard type of radars and other visual and electronic navigation aids.

The current surveillance utilities for receiving Radar and VHF Voice Position Reports (VPR) and HF Radio Data/VPR between ships and TCC and Maritime Traffic Management (MTM) can be detected by Radar and MF/HF/VHF CRS. This subsystem may have similar propagation problems and limited range or when ships are sailing inside of fiords and behind high mountains Coastal Radar cannot detect them; see the Surveillance Subsystem in Figure 4. The very bad weather conditions, deep clouds and heavy rain could block radar signals totally and on the screen will be blanc picture without any reflected signals, so in this case cannot be visible surrounded obstacles or traffic of ships in the vicinity, and the navigation situation is becoming very critical and dangerous causing collisions and huge disasters [8, 9, 10].

On the contrary, the new Communication CNS/MTM System utilizes the communications satellite and it will eliminate the possibility of interference by very high mountains, see all three CNS Subsystems in Figure 4.

At this point, satellite voice communications, including a data link, augments a range and improves both the quality and capacity of communications. The WX and NX warnings, sailing planning and NAVAREA information may also be directly input to the Navigation Management System (NMS).

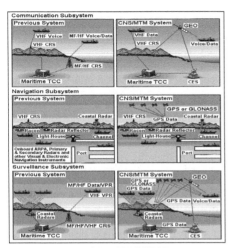

Figure 4. Current and New CNS/MTM System
Courtesy of Book: "Global Mobile Satellite Communications" by Ilcev [6]

The new Navigation CNS/MTM System is providing improved GPS/GLONASS navigation data, while Surveillance CNS/MTM System is utilizing augmented facilities of GPS or GLONASS signals. Thus, if the navigation course is free of islands or shallow waters, the GPS Navigation Subsystem data provides a direct approaching line and the surveillance information cannot be interfered by mountainous terrain or bad weather conditions. The display on the screen will eliminate misunderstandings between controllers and ship's Masters or Pilots [02, 10].

4 MARITIME MOBILE SATELLITE SERVICE (MMSS)

The MMSS functions in frame of the new MTAS infrastructure include the provision of all the mobile maritime communications defined by the IMO, such as new Global Maritime Distress and Safety System (GMDSS), Inmarsat and Cospas-Sarsat systems, including new systems with nomenclatures such as Maritime Commercial Communications (MCC) and Maritime Crew and Passenger Communications (MCPC).

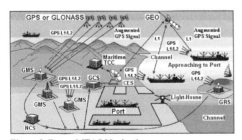

Figure 5. Future MTAS Navigation

In a more general sense, these MSC service solutions could be available for MTM, Maritime Traffic Control (MTC) and Maritime Traffic Service (MTS) providers and maritime operators in all ocean regions through data link service providers. Direct access to the MTAS network could also be possible through the implementation of dedicated GES in other states covered by MTAS spacecraft.

The MTAS system for the SES is interoperable with MSC system of the Inmarsat Space and Ground network. It can be connected directly to the navigation bridge GMDSS operator (Master, duty-deck or radio officer) by VDV, Fax, video, GPS augmentation information and Automatic Dependent Surveillance System (ADSS).

The MTAS will not only be capable of handling MTS for ocean going vessels, but will also be offered to the Civil Maritime Community (CMC) in all coastal regions as an infrastructure, which could facilitate the implementation of the future IMO CNS/MTM systems.

The MTAS service provides all ocean going vessels with GPS augmentation information to improve safety and security at sea and all navigational performance requirements, namely to find out the response to the demands of ICAA, which are essential to the use of GPS or GLONASS for vessels operation as the sole means of navigation. Using previous not augmented system, ship navigation officers know very well where their ship is in space and time, but offshore MTC terminals don't know. In order to provide all ships and MTC with sufficient GPS augmentation information and satellite surveillance, a certain number and location of GMS will be required. At this point, the number and location of GMS required for each state in the region will depend on the requirements for the level of navigation services and reception of GPS signals. The MTAS system needs number of GMS, few GCS and GES for the each region [6, 10].

4.1 Current Radio and MSC System

The previous Maritime Radio Communications (MRC) system for general international purposes has been operational over 100 years and recently was replaced by MSC system to enhance ship-to-shore voice and data traffic for both commercial and safety applications. In general, the initial development will have been established by using a service of MRC on MF and HF Morse radiotelegraphy, radio telex and radiotelephony (voice) for maritime medium and long distance communications, respectively. The latter progress was in order to promote advanced maritime short distance commercial, safety, approaching and on scene distress communications on VHF voice frequency band. Finally, global DCS MF/HF/VHF Radio subsystem was developed by IMO in frame of

GMDSS system and integrated with Inmarsat and Cospas-Sarsat facilities.

Meanwhile, in order to respond to the significant increase in the volume of communications data that has accompanied the large increases in cargo maritime traffic, periodic communications have moved to the satellite communications low, medium and high speed data link and data transmission has become the core type of maritime communications. The media needs to be divided to reflect this change in communications content, which has seen voice (Tel) communications used mainly for irregular safety and security or even for emergency situations in general. A transmission system based on fundaments new GMDSS digital technology (bit-based) needs to be integrated by the MTAS, to introduce wholesale improvements in Satellite CNS ability and to enhance current system for emergency (distress, safety and security) [06, 10].

Gradually, new MMSS VDV and VDVoIP links have come into use and totally may replace old HF and VHF traditional radio. Because of any emergency and very bad weather conditions ship can be extremely affected, it is necessary to keep them as alternative solutions and to employ again a well-trained Radio Officer on board every oceangoing ship. However, in normal circumstances and for fast communication impact SES can be used for communications with corresponding GES via any MATS or Inmarsat GEO satellite for maritime commercial, emergency and social purposes [6].

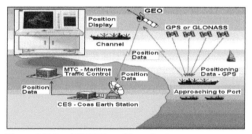

Figure 6. Future MTAS Surveillance System
Courtesy of Book: "Global Mobile Satellite Communications" by Ilcev [6]

4.2 Integration of RSAS and GNSS

The GPS or GLONASS can be used worldwide to control the positions of vessels and to manage maritime traffic for oceangoing and coastal navigation. They support vessel's navigation well in all routing phases, including approaching to the port and mooring utilities. In fact, they have some performance limitations and they cannot consistently provide the highly precise and quite safe information in the stable manner required for wide-area navigation services. To assure safe and efficient sea traffic navigation of civil vessels, GPS and GLONASS performance needs to be augmented with another

system that provides ICAA essential elements well for sea navigation. The MTAS augmentation solution for GPS/GLONASS can be integrated with adequate Land Transportation Augmentation System (LTAS) and Aeronautical Transportation Augmentation System (ATAS) into the US WAAS, Japanese MSAS, European EGNOS, Russian SDCM, Chinese SNAS, Indian GAGAN and new systems such as ASAS, Australian and South American RSAS. Once in operation, this new state-of-the-art system will assure full navigation services for vessels in all navigation phases within the oceanwide, coastal, approaching and channel waters through GSAS coverage.

The L1/L2 RF band is nominated for the transmission of signals from GNSS spacecraft in ground and air directions, which can be detected by the GMS, GES and GNSS-1 onboard ship's receivers. Otherwise, the MTAS GNSS satellite transponder uses the L1 RF band to broadcast GNSS augmentation signals in the direction from GES to SES. The L, Ku or Ka-band is used for unlinking GNSS augmentation data from SES via GEO spacecraft to TCC. The whole ground infrastructure and Communication System is controlled by GCS and Network Control System (NCS). The components of the MTAS navigation system are illustrated in Figure 5. To provide GNSS augmentation information, all ground stations, which monitor GNSS signals, are necessary in addition to MTAS. This special navigation infrastructure, which is composed by MTAS, GPS/GLONASS or GNSS wide-area augmentation system and these ground stations, is called the MTAS network [02, 06].

4.3 Wide Area Navigation (WANAV) System

The Wide Area Navigation (WANAV) system is a way of calculating own precise position using the Ship Surveillance Satellite Equipment (SSSE) facilities and other installed onboard ship navigation devices to navigate the desired course and to send this position to TCC. In the case of WANAV routes it has been possible to connect in an almost straight line to any desired point within the area covered by the satellite equipment and service.

In any event, setting the WANAV routes has made it possible to ease congestion on the main sea routes and has created double tracks. This system enables more secure, safety and economical sea navigation routes.

4.4 MTAS Automatic Dependent Surveillance System (ADSS)

The current radio surveillance system is mainly supported by VHF CRS. Namely, this system enables display of real-time positions of the nearby approaching ships using radar and VHF voice radio equipment. Due to its limitations, the VHF service being used for domestic sea space, channels and coastal waters cannot be provided over the ocean. Meanwhile, out of radar range and VHF coverage on the oceanic routes, the ship position can be regularly reported by HF radio voice or via data terminals to the HF CRS.

Consequently, the advanced CNS/MTM system utilizes the ADSS data function, which automatically reports all current ships positions measured by GPS to MTC, as illustrated in Figure 6. In this way, the approaching vessels receives positioning data from GPS spacecraft or GPS augmented data via GEO satellite transponder, as illustrated in Figure 5, and then sends via GES its current position for recording and processing to the MTC terminal and displaying on the like radar screen. This service enhances safety, security and control of vessels in ocean and coastal navigation.

The screen display of satellite ADSS looks just like a pseudo-radar coverage picture showing positions of the ships. The new ADSS system will increase safety and security at sea and reduce ships separation, improve functions and selection of the optimum route with more economical courses. It will also increase the accuracy of each ship position and reduce the workload of both controller and ship's Master or Pilot, which will improve safety and security. In this sense, ships can be operated in a more efficient manner and furthermore, since the areas where VHF radio does not reach due to the short range, mountainous terrain or bad weather will disappear, small ships, including Pilot boats and helicopters, will be able to obtain any data and safety information on a regular basis. These functions are mandatory to expand the traffic capacity of the entire ocean or coastal regions for all ships and for the optimum navigation and safety route selection under limited space and time restraints [02, 06].

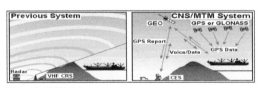

Figure 7. SESR Subsystem
Courtesy of Paper, "Satellite CNS for MTAS" by St. D. Ilcev [2]

5 SPECIAL EFFECTS OF THE MTAS SYSTEM

Special effects of the MTAS system used for secure communications, navigation, ranging, logistics and control of the vessels at sea, in the channels, around the coastal waters and in the port surface ship traffic are Safety Enhancements on Short and Long Ranges, Reduction of Separation Minima, Flexible Sailing Profile Planning and Coastal Movement Guidance and Control.

These effects of the MTAS are very important to improve maritime communication facilities in any phase of sailing, to enable better control of ships, provide flexible and economic trip with optimum routes, to enhance surface guidance and control in port and in any case to improve safety and security at sea and in the ports.

5.1 Safety Enhancements at Short and Long Ranges

A very important effect of the new MTAS system for CNS/MTM is to provide Safety Enhancement at Short Ranges (SESR) via GES, as illustrated in Figure 7.

Current radio system for short distances between vessels and CRS is provided by VHF voice or by new DSC VHF voice and data equipment, so the ship's Master or Pilot will have many problems establishing voice bridge radio communications when the ship position is in the shadow of high mountains in coastal waters.

Meanwhile, all vessels sailing in coastal waters or fiords and in ports can receive satellite navigation and communications even at short distances and where there is no navigation and communications coverage due to mountainous terrain. This is very important for safety and secure navigation during bad weather conditions and reduced visibility in channels, approachings and coastal waters, to avoid collisions and disasters.

The MTAS system is also able to provide Safety Enhancement at Long Rangers (SELR) illustrated in Figure 8, by using faded HF radio system or the noise-free satellite system. In such a way, many ships out of VHF range can provide their augmented or not augmented positions to MTC or will be able to receive safety and weather information for secure navigation [02, 06].

5.2 Reduction of Separation Minima (RSM)

One of the greatly important MTAS safety navigation effects is the Reduction of Separation Minima (RSM) between ships or other moving object on the sea routes by almost half, as shown in Figure 9. The current system has an RSM controlled by conventional VHF or HF Radio system and Radar Control System (RCS), which allows only large distances between vessels. However, the new CNS/MTM system controls and ranges greater numbers of vessels for the same sea corridors (channels), which enables minimum secure separations, with a doubled capacity for vessels and enhancements of safety and security. Therefore, a significant RSM for sailing ships will be available with the widespread introduction and implementation worldwide of the new RSAS technologies on the CNS system [6, 7].

Figure 8. SELR Subsystem
Courtesy of Paper, "Satellite CNS for MTAS" by St. D. Ilcev [2]

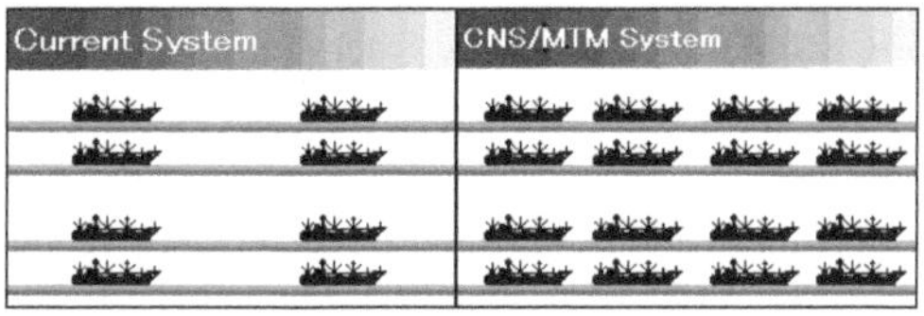

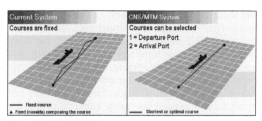

Figure 9. RSM Subsystem
Courtesy of Book, "Maritime CNS" by St. D. Ilcev [4]

5.3 *Flexible Sailing Profile Planning (FSPP)*

The next positive effect of MTAS system is Flexible Sailing Profile Planning (FSPP) of shortest or optimal course, shown in Figure 10. The current system uses fixed courses of orthodrome, loxodrome and combined navigation by navaids. Thus, the fixed course is controlled by the vessel's on-board navigation instruments only, which is a composite and not the shortest possible route from departure to arrival at the destination port. The FSPP allows the selection of the shortest or optimum course between two ports and several sub points. With thanks to new RSAS technologies on CNS/MTM system FSPP will be available for more economic and efficient sailing operations. This means that the ship's engines will use less fuel by selecting the shortest sailing route of new CNS/MTM system than by selected the fixed courses of current route composition [6, 7].

6 LSAS SYSTEM CONFIGURATION

The LSAS system configuration is intended to complement the CNS service for local environment of seaport using a single differential correction that accounts for all expected common errors between a local reference and mobile users. The LSAS infrastructure will broadcast navigation information in a localized volume area of seaports or airports using satellite service of satellite CSN solutions or any of mentioned RSAS networks developed in Northern Hemisphere.

As stated earlier, any hypothetical RSAS network will consist a number of GMS (Reference Stations), several GCS (Master Stations) and enough GES (Gateways), which service has to cover entire mo-

bile environment of dedicated region as an integrated part of GSAS. Inside of this coverage the RSAS network will also serve to any other customers at sea, on the ground and in the air users, who needs very precise determinations and positioning, such as:

1 Maritime (Shipborne Navigation and Surveillance, Seafloor Mapping and Seismic Surveying);
2 Land (Vehicleborne Navigation, Transit, Tracking and Surveillance, Transportation Steering and Cranes);
3 Aeronautical (Airborne Navigation and Surveillance and Mapping);
4 Agricultural (Forestry, Farming and Machine Control and Monitoring);
5 Industrial, Mining and Civil Engineering;
6 Structural Deformations Monitoring;
7 Meteorological, Cadastral and Seismic Surveying; and
8 Government/Military Determination and Surveillance (Police, Intelligent services, Firefighting); etc.

In a more general sense, all above fixed or mobile applications will be able to assess CNS service inside of RSAS coverage directly by installing new equipment known as augmented GPS or GLONASS Rx terminals, and so to use more accurate positioning and determination data.

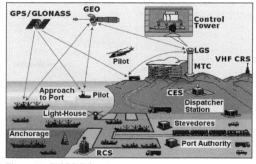

Figure 10. FSPP Subsystem
Courtesy of Book, "Maritime CNS" by St. D. Ilcev [4]

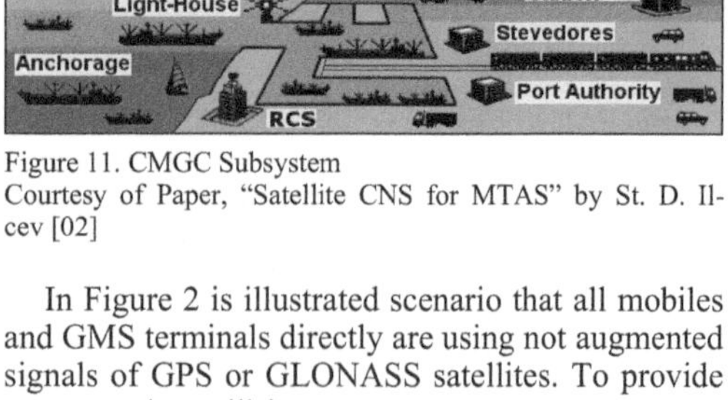

Figure 11. CMGC Subsystem
Courtesy of Paper, "Satellite CNS for MTAS" by St. D. Ilcev [02]

In Figure 2 is illustrated scenario that all mobiles and GMS terminals directly are using not augmented signals of GPS or GLONASS satellites. To provide augmentation will be necessary to process not augmented signals in GCS, to eliminate all errors and produce augmented signals. However, in this stage any RSAS network standalone will be not able to produce augmented service for seaports, airports or any ground infrastructures.

At this point, it will be necessary to be established some new infrastructure known as an LSAS, which can provide service for collecting augmented data from ships, land vehicles, airplanes or any ground user. The navigation data of mobiles can be processed in the TCC cites and shown on the surveillance screen similar to the radar display and can be used for traffic control system at the see, on the ground and in the air. This scenario will be more important for establishment MTC or Air Traffic Control (ATC) service using augmented GNSS-1 signals from the ships or aircraft, respectively. In this sense, the LSAS network can be utilized for seaports known as Coastal Movement Guidance and Control (CMGC) and airports as Surface Movement Guidance and Control (SMGC) [04, 10].

7 COASTAL MOVEMENT GUIDANCE AND CONTROL (CMGC)

The new LSAS network can be implemented as a Coastal Movement Guidance and Control (CMGC) system integrated in the CNS of any RSAS infrastructure. It is a special maritime security and control system that enables a port controller from Control Tower at shore to collect all navigation and determination data from all ships and vehicles, to process these signals and display on the surveillance screens. On the surveillance screen can be visible positions and courses of all ships in vicinity sailing areas, so they can be controlled, informed and managed by traffic controllers in any real time and space [6].

In this case, the LSAS traffic controller provide essential control, traffic management, guide and monitor all vessels movements in coastal navigation, in the cramped channel strips and fiords, approaching areas to the anchorage and harbours, ship movement in the harbours, including land vehicles in port and around the port's coastal environment, even in poor visibility conditions at an approaching to the port. The controller issues instructions to the ship's Masters and Pilots with reference to a command surveillance display in a Control Tower that gives all vessels position information in the vicinity detected via satellites and by sensors on the ground, shown in Figure 11.

The command monitor also displays reported position data of coming or departing vessels and all auxiliary land vehicles (road and railways) moving into the port's surface. This position is measured by GNSS, using data from GPS/GLONASS and GEO satellite constellation. A controller is also able to show the correct ship course to Masters and sea Pilots under bad weather conditions and poor visibility or to give information on routes and separation to other vessels in progress. The following segments of CMGC infrastructure are illustrated in Figure 11:

1 GPS or GLONASS GNSS Satellite measures the vessel or port vehicle's exact position.
2 GEO MSC Satellite is integrated with the GPS positioning data network caring both communication and navigation payloads, In addition to complementing the GPS satellite, it also has the feature of communicating data between the ships or vehicles and the ground facilities, pinpointing the mobile's exact position.
3 Control Tower is the centre for monitoring the traffic situation on the channel strips, approaching areas, in the port and around the port's coastal surface. The location of each vessel and ground vehicle is displayed on the command monitor of the port control tower. The controller performs sea-controlled distance guidance and movements for the vessels and ground-controlled distance vehicles and directions based on this data.
4 Light Guidance System (LGS) is managed by the controller who gives green light or red light guidance whether the ship should proceed or not by pilot in port, respectively.
5 Radar Control Station (RCS) is a part of previous system for MTC of ship movement in the channels, approaching areas, in port and around the port's coastal environment.
6 Very High Frequency (VHF) is Coast Radio Station (CRS) is a part of RCS and VHF or Digital Selective Call (DSC) VHF Radio communications system.
7 Ground Earth Station (GES) is a main part of satellite communications system between GES terminals and shore telecommunication facilities via GEO satellite constellation.
8 Pilot is small boat or helicopter carrying the special trained man known as a Pilot, who has to proceed the vessel to the anchorage, in port, out of port or through the channels and rivers.
9 Bridge Instrument of each vessel displays the ship position and course [02, 06].

8 CONCLUSION

The CNS has been set up to identify the possible applications for global radio and satellite CNS, safety and security and control of aircraft, freight and passengers and SAR service in accordance with IMO and ICAO regulations and recommendations. The new satellite CNS using GEO satellites with Communication and GNSS payloads for MTC/MTM is designed to assist navigation both sailing at open sea and approaching to the anchorages and seaports. The potential benefits will assist MTC to cope with increased maritime traffic and to improve safety and reducing the infrastructures needed at shore. The Communication payloads usually at present employ transponders working on RF of L/C, Ku and recently on Ka-bands for DVB-RCS scenario. Because that Ku-band is experiencing some transmission problems and is not so cost effective, there is proposal that Ka-band will substitute Ku-band even in mobile applications including maritime and aviation.

When planning maritime routes and berthing schedules at busy seaports, it is essential to ensure that ships are always at safe distance from each other and that they are passing some critical channels safely. The trouble is that it is not always possible to figure out where the ships are, especially during very bad weather conditions. It is necessary to reduce the margins of critical navigation and increase the safety of ships in each sea and passage corridors. The new CNS GNSS-1 networks of MTAS and forthcoming European Galileo and Chinese Compass GNSS-2 will provide a guaranteed service with sufficient accuracy to allow ship's masters and pilots including MTC to indicate a current position and safety margins reliably and precisely enough to make substantial efficiency sailing. The GNSS helps masters to navigate safely, especially in poor weather conditions and dense fog, in which sailing using CNS via RSAS system or DGPS is reliable. Any seaports are unlikely to invest in this system, but they can use CNS of global or local augmented system or when Galileo and Compass become operational, the need for a differential antenna will reduce costs. Galileo and Compass will also need implementation of CNS via RSAS, so their guaranteed service and use of dual frequencies will increase accuracy and reliability to such an extent that vessels will be able to use safely their navigational data for guidance including their on-board technology alone.

REFERENCES

[01] Ilcev D. St. "Global Aeronautical Communications, Navigation and Surveillance (CNS)", John Wiley, Chichester, 2010.

[02] Ilcev D. St. "Satellite CNS for Maritime Transportation Augmentation System (MTAS)", CriMiCo Conference, IEEE Catalog Numbers CFP09788, Sevastopol, Ukraine 2009.

[03] Kaplan E.D. "Understanding GPS Principles and Applications", Artech House, Boston, 1996.

[04] Ilcev D. St. "Maritime Communications, Navigation and Surveillance (CNS)", DUT, Durban, 2010.

[05] Group of Authors, "Website of EGNOS (www.esa.int), WAAS (www.gps.faa.gov) GSAS", 2008.

[06] Ilcev D. St. "Global Mobile Satellite Communications for Maritime, Land and Aeronautical Applications", Springer, Boston, 2005.

[07] Group of Authors, "MTSAT Update", NextSAT/10 CG, Japan Civil Aviation Bureau, MSAS, Tokyo, 2009.

[08] El-Rabbany A. "Introduction to GPS", Artech House, Boston, 2002.

[09] Grewal M.S. and others, "Global Positioning Systems, Inertial Navigation, and Integration", Wiley, London, 2008.

[10] Prasad R. & Ruggieri M., "Applied Satellite Navigation Using GPS, GALILEO, and Augmentation Systems", Artech House, Boston, 2005.

14. On a Data Fusion Model of the Navigation and Communication Systems of a Ship

G. K. Park & Y.-K. Kim

Division of Maritime Transportation System. Mokpo National Maritime University, Republic of Korea

ABSTRACT: Ship mates should be aware of images, numerical values, texts and audio-based information of radar, AIS, NAVTEX, VHF, and etc. for safe navigation. However, it is very complicated and difficult for them to acquire such information and use it as data for decision-making for safe navigation while keeping watch for navigation. So, a system to understand, unite and provide multimedia marine information for mates in voice is necessary. This study tries to suggest data fusion model of the navigation and communication system.

1 INSTRUCTIONS

A ship's navigation instrument produces and provides a variety of information necessary for safe navigation.

A number of researches' efforts to provide more advanced information for safe navigation.

Related studies include a study on collision avoidance assistant system using fuzzy case-based reasoning[1] and a study on a conceptual model for collision avoidance for ontology-based fuzzy CBR support system[2], and a study on a system that automatically sets up routes and supports the description of the set route for mates on small ships, who are lacking of expertise[3]

In addition, there are a study that built an embedded system that can control ships steering system of ships with speech wheel order[4], speech-recognition-based intelligent ships steering control system suggested by Gyei-Kark Park and Ki-Yeol Seo[5], and a study on ontology based fuzzy ships steering control system[6].

However, studies so far have been carried on systems for navigators without expertise on small ships without automatic control, and related systems or studies which can provide comprehensive information efficiently for experts are lacking.

This study tries to suggest data fusion models necessary for the construction of the system to understand, unite, and provide the multimedia marine information for mates. A ship's navigation instrument produces and provides a variety of information necessary for safe navigation.

2 OVERVIEW OF A DATA FUSION MODEL FOR NAVIGATION INFORMATION

Such a process as Figure 1 is necessary for establishing a data fusion model to recognize a situation using the multimedia navigation safety information provided by diverse navigation equipment, induce the information needed for decision-making, and provide it linguistically.

First of all, this paper composes a data field by navigation equipment by analyzing the raw data produced and provided by navigation equipment such as GPS, ARPA, AIS, NAVTEX, VHF, etc. and then establishes a knowledge representation model to express the data in each data field and the relationship between objects as subjects between attributes linguistically using a semantic network.

Next, this paper establishes a data fusion model using a knowledge representation model, and provides the information obtained newly by providing the information of a data fusion model in a language or by fusing or inducing the information provided by a data fusion model in a language.

Figure 1. Diagram of providing the inferred information

3 KNOWLEDGE REPRESENTATION AND DATA FUSION MODEL

3.1 *Data field of navigational equipments*

To use the information provided by navigation equipments, the information provided by the equipment was analyzed and data field for the equipment were prepared.

GPS, ARPA and AIS information should provide which is specified in SOLAS resolutions.

NAVTEX provides information received in the unit of a character of English alphabet in a text form, natural language processing is completed so it was assumed that an appropriate data field was obtained.

Most navigation warning, weather warning and other urgent safety-related notices NAVTEX provides were discovery of a new navigation obstruction, change of aids to navigation, construction or training section, or threatening weather occurrence, etc., and such information consists of the name, the term of validity and location or area.

VHF is communication equipment using frequency. Since it communicates using human language directly, information and types obtained using this equipment are unlimited. However, the most important thing is that it has a merit that through which the intention of the other ship can be known.

Using these, data fields were prepared.

To use information that can be provided by navigational equipments, it was assumed that speech recognition and natural language processing had been completed.

3.2 *Knowledge Representation Models*

The Knowledge Representation Model by navigation instruments was built by expressing the knowledge relationship between the data related to the subject and the subject using semantic network, and the information which each equipment provides was expressed in simple sentences, and the knowledge representation by the subjects was constructed using the knowledge representation by navigation instrument.

3.2.1 *Knowledge Representation Model of GPS*

All information provided by GPS is included.

Figure 2 is Knowledge Representation Model of GPS and it expressed information which is provided by this model in a simple sentence.

"Ownship's Position's GPS position Is latitude 34° 12.5'N, longitude 126° 22.4'E."

"Ownship's Speed Is 18.3 kts."

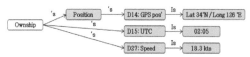

Figure 2. Knowledge Representation Model of GPS

3.2.2 *Knowledge Representation Model of ARPA*

All the information provided by ARPA is included and D1: Bearing & D2: Range as location information added vertices as position before connecting with subject for amalgamation with information which is provided by other navigation instrument.

Figure 3 is Knowledge Representation Model of ARPA, and it expressed information which is provided by this model in a simple sentence.

"Ship 1's Position is Bearing 312 degree."

"Object 1's CPA Is 0.0 miles."

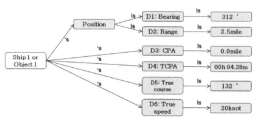

Figure 3. Knowledge Representation Model of ARPA

3.2.3 *Knowledge Representation Model of AIS*

All information which is provided by AIS can be expressed but only D14: GPS position as location information was connected to ship as subject through the vertices as position.

Figure 4 is Knowledge Representation Model of AIS, and it expressed information which is provided by this model in a simple sentence.

"Ship 1's Heading Is 000 degree."

"Ship 1's Name is SAEYUDAL."

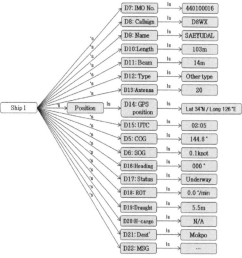

Figure 3. Knowledge Representation Model of AIS

3.2.4 Knowledge Representation Model of NAVTEX

It was designed using data field of NAVTEX.

Figure 5 is Knowledge Representation Model of NAVTEX, and it expressed information which is provided by this model in a simple sentence.

"Object 1's Name is Dangerous wreck."

"Object 1's Position is latitude 34° 12.5' N and longitude 126° 22.5' E."

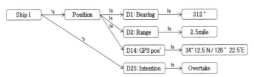

Figure 5. Knowledge Representation Model of NAVTEX

3.2.5 Knowledge Representation Model of VHF

It was designed using data field of VHF.

Figure 6 is Knowledge Representation Model of VHF, and it expressed information which is provided by this model in a simple sentence.

"Ship 1's Intention Is overtake."

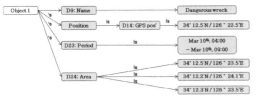

Figure 6. Knowledge Representation Model of VHF

3.3 Data fusion algorithm

A data fusion process is necessary for fusing a knowledge representation model into a data fusion model.

In order to judge that the data provided by two knowledge representation models are the information of the same objects, the data with the same meaning among the data provided by two knowledge representation models should be comparable, and it can be judged that the data provided by two knowledge representation models are the information with the same objects when their similarity is within a certain range after comparison.

Figure 7 shows a proposed data fusion algorism.

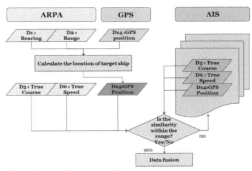

Figure 7. Data fusion algorithm

3.4 Data fusion model

3.4.1 Data fusion model in the case of a ship as target

In the case of ship as subject, information can be obtained using ARPA, AIS and VHF in general. Therefore, it was represented by combining knowledge representation models of ARPA, AIS and VHF.

Figure 8 is data fusion model in the case of a ship as target.

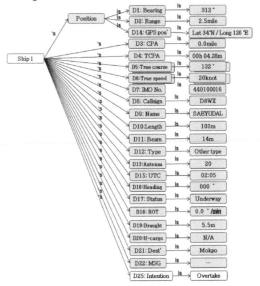

Figure 8. Data fusion model in the case of a ship as target

3.4.2 Data fusion model in the case of a object as target

In the case of subject as except ship, the information can be obtained by ARPA or NAVTEX etc, so it was represented by combining knowledge representation models of ARPA and NAVTEX.

Figure 9 is Data fusion model in the case of a object as target.

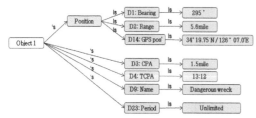

Figure 9. Data fusion model in the case of the subject as except ship

4 APPLICATION OF DATA FUSION MODEL

4.1 Description of navigation situation using data fusion model by subject

It explained the navigation situation and the information of other ships and marine obstacle which mate can obtain in the given navigation situation was expressed using Data Fusion Model by subject in the sentence.

The navigation situation of Figure 10 is a dangerous one in which two ships are encountering while they pass through a narrow channel.

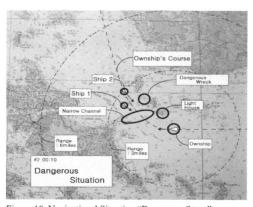

Figure 10. Navigational Situation "Dangerous Stage"

4.2 Navigation situation expression of information that can be obtained from Data Fusion Model

Figure 11 shows a data fusion model for the information acquirable in a dangerous situation of scenario.

The data fusion model in Figure 11 can provide all the information provided by navigation equipment in a navigation of scenario in a simple sentence.

This expresses the meaning of information on Ship1, Ship2 and Object1 provided by a given data fusion model in a simple sentence.

"Ship 1's Position Is D1:Bearing Is 304°."

"Ship 1's Position Is D2:Range Is 2.12 miles."

"Ship 1's Position Is D14:GPS position Is Latitude 34° 19.4´ N and longitude 126° 05.85´ E.

"Ship 1's D3:CPA Is 0.15 miles."

"Ship 1's D4:TCPA Is 5.2 minute."

"Ship 2's D5:True course Is 126°."

"Ship 2's D6:Ture speed Is 10.0 kts."

"Ship 2's D7:IMO No. Is 440100001."

"Ship 2's D18:Rate of turn Is 0.0°/min."

"Object 1's Position Is D1:Bearing Is 332°."

"Object 1's Position Is D2:Range Is 1.7 miles."

"Object 1's Position Is D14:GPS position Is Latitude 34° 19.75´ N and longitude 126° 07.0´ E.

"Object 1's D9:Name Is Dangerous wreck."

"Object 1's D23:Period Is Unlimited."

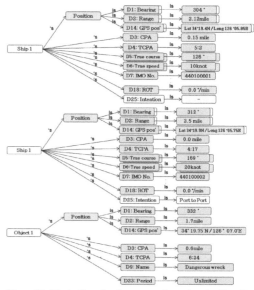

Figure 11. Navigation situation expression using the Data Fusion Model for each object

4.3 Navigation situation expression using the Data fusion model

Figures 12, 13 and 14 express the information obtainable through induction by a mate in a navigation situation of scenario using a data fusion model in Figure 11.

Figure 12 expresses the information of a new meaning inducing sentence using the meaning of multiple objects "the CPA of Ship1, Ship2 and Object1 is all within 1mile and less than 7 minutes, so it is very dangerous" using a data fusion model.

The meaning except for the meaning created newly after induction can be expressed linguistically.

- Ship1`s D3, D4, Ship2`s D3, D4, Object1`s D3, D4.

"Ship1's CPA Is 0.15mile, TCPA Is 5:2, Ship2`s CPA Is 0.0mile, TCPA Is 4:17, Object1`s CPA Is 0.6mile, TCPA Is 6:24."

Figure 12. Navigation situation expression – 1

Figure 13 expresses the information of a sentence induced by combining the meaning of an object acquired from multiple navigation equipment "Object1 has Bearing 332°, Range 1.7mile, CPA 0.6mile, and TCPA 6:24, and its name is Dangerous wreck"using a data fusion model.

All the meanings of a sentence can be expressed.
-Object1`s D1, D2, D3, D4, D9.

"Object1 `s Bearing Is 332°, Range Is 1.7mile, CPA Is 0.6mile, TCPAIs 6:24, Name Is Dangerous wreck."

Figure 13. Navigation situation expression - 2

Figure 14 expresses the information of a sentence inducing a situation and presenting a solution by combining the meanings of an object acquired from multiple navigation equipment "Ship 1 has CPA 0.15mile and TCPA 5min 2sec, but it doesn't veer and respond to communication, so do DSC(Digital Selective Calling) for IMO No. 440100002"using a data fusion model.

The induced meaning cannot be expressed, but the information can be expressed.

- Ship1`s D3, D4, D18, D25, D7.

"Ship1`s CPA Is 0.15mile, TCPA Is 5min 2sec, Rate of turn Is 0.0°/min, Intention is nothing, DSC Is 440100002."

Ship1`s CPA Is 0.15mile, TCPA Is 5min 2sec, But, Rate of turn Is 0.0°/min, Intention is nothing ∴ DSC 440100002.

Ship1`s D3, D4 But D18, D25, DSC D7.

Figure 14. Navigation situation expression - 3

5 CONCLUSIONS

This study proposed a data fusion model using semantic network to analyze the information provided by each navigation equipment and express the meaning of information provided by navigation equipment by selecting navigation equipment such as GPS, ARPA, AIS, NAVTEX, VHF receiver, etc. providing essential information for grasping a navigation situation, and explained the information acquirable and the inducible by a mate in a navigation situation of scenario using a data fusion model.

The test of the proposed model in some real navigation situations will be done to verify its validity in future.

REFERENCES

Gyei-Kark Park, John Leslie Benedictos, "Ship Collision Avoidance Support System Using Fuzzy-CBR", Korean Institute of Intelligent System, Journal of Korean Institute of Intelligent System, Vol. 16, No. 5, pp. 635-641, 2006.

Gyei-Kark Park, Woong-Gyu Kim, John Leslie RM Benedictos, "Conceptual Model for Fuzzy-CBR Support System for Collision Avoidance at Sea Using Ontology", Korean Institute of Intelligent System, Journal of Korean Institute of Intelligent System, Vol. 17, No.3, pp. 390-396, 2007.

Tae-ho Hong, Ki-Yeol Seo, Gyei-Kark Park, "Building of an Integrated Navigation Guiding System Based on ENC", Korean Institute of Intelligent System, Proceedings of KIIS Spring Conference, Vol. 15, No.1, pp. 394-399, 2005.

Gyei-Kark Park, Ki-Yeol Seo, Tae-ho Hong, "Development of an Embedded System for Ship's Steering Gear using Voice Recognition Module", Korean Institute of Intelligent System, Journal of Korean Institute of Intelligent System, Vol.14, No.5, pp.604-609, 2004.

Ki-Yeol Seo, Se-Woong Oh, Sang-Hyun Suh, Gyei-Kark Park, "Intelligent Steering Control System Based On Voice Instructions", International Journal of Control Automation and System, Vol. 5, No.5, 10, pp. 539-546, 2007.

Ki-Yeol Seo, Gyei-Kark Park, Chang-Shing Lee, Mei-Hui Wang, "Ontology-based Fuzzy support agent for ship steering control", Expert systems with applications, Vol 36, Issue 1, pp. 755-765, 2009.

15. Automation of Message Interchange Process in Maritime Transport

Z. Pietrzykowski, G. Hołowiński, J. Magaj & J. Chomski
Maritime University of Szczecin, Szczecin, Poland

ABSTRACT: The paper is focused on automation of data message interchange in maritime transport. A general concept of communication system is proposed. The authors deal with some issues of automatic communication in marine navigation. The principles and form of communication based on the use of maritime transport communication ontology with XML Schema are described.

1 INTRODUCTION

Some shipboard systems and equipment (AIS, GMDSS, ARPA, Navtex, and GPS) are used in the automation of information acquisition and exchange. However, these systems do not ensure exchange of information in complex situations, where co-operation between navigators (or coast station officers, helicopter pilots, etc.) has to be established.

The automation of the message interchange process in maritime transport could support navigators in this case. Moreover, such automation is a basis for further development of more complex, agent based navigation support system including an automated negotiation layer. Such automated negotiation systems are well known in e-business and trading environments and presented, among others, in Beam 1997, Paurobally 2003, Karp 2004.

Herein proposed is an approach to solve the automation of the message interchange process in maritime transport. This paper shows results of the research continued after the one described in Pietrzykowski et al. 2003, 2005, 2006. A general concept of communication system is shown. The ontological structure of messages is introduced and its description in XML Schema is proposed to formalize the format of message contents and is an extension of the navigational based ontology in Malyankar et al. 1999, Mingyang P. et al. 2003, Kopacz et al. 2004 and Pietrzykowski et al. 2006.

The proposed solutions are based on an analysis of a real process of communication between navigators presented as example in Pietrzykowski et al. 2006.

2 A CONCEPT OF AUTOMATION OF MESSAGE INTERCHANGE

The transformation of communication from human-to-human to fully automated one is a continuous process. Its purpose is not to provide the environment for fully automated communication between machines, but rather to allow communicating between:

– humans (system operators),
– machines (e.g. exchanging information between ships),
– humans and machines (in all possible combinations and proportions).

Figure 1 shows the scheme of the proposed communication between the sender and the receiver. A message built on sender's side can include information from the operator (e.g. navigator), even if their primary source is any of the available electronic systems. This information - sentences - can be put in manually by the operator. Besides, some information contained in a message is taken directly from external electronic systems (e.g. shipboard AIS). The aim of the communication system is to compose a valid message by means of the previously defined commonly understood syntax using the information contained in these sentences.

The receiver's system should be able to read this message and decompose it to small sentences shown directly to the operator, to store it or send to any of the available external systems. Moreover, the operator can obtain additional information from the external systems after they process any data from the message.

The proposed concept is a technical communication basis for further research in the following areas:

– visualization of information,

- automated negotiations between objects of communication in maritime transport,
- the latest systems on the market show the important role of efficient and ergonomic interface (e.g. visual interfaces of mobile devices). The interface for communication system in maritime transport should support the visualization of both source and destination objects of the communication. It should ensure that navigators understand who takes part in this process (visual verification of participants of the communication is provided as support for the operator).

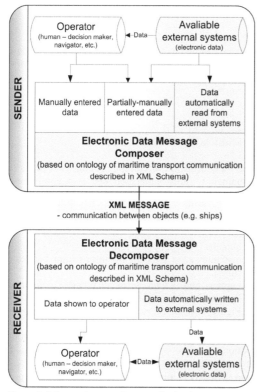

Figure 1. Scheme of proposed communication between two objects (e.g. ships). Note: a gray marked elements show automated data processing.

The proposed system concept can be regarded as a basis for an automated negotiation system in maritime transport, which can be used, for instance, as an expert system helping to optimize navigation issues. In that case the automated or semi-automated negotiation support system requires at least the following functionalities: information sharing among objects (e.g. ship positions, speeds, courses) and auto-negotiation between them (e.g. implemented as agents).

In Beam 1997 it is acknowledged, with a support of several examples, that building an automated negotiation system is a challenging and difficult task.

Both the need for ontology and the need for a negotiation strategy are highlighted that study. Ontology is a way of categorizing objects in such a way that they are semantically meaningful to a software agent. The negotiation strategy in the navigational environment should be clear in maritime transport (while the strategy is a secret in known trading systems).

However, the functionalities indicated above cannot be realized without automated communication. The sections that follow present an ontology and its implementation in XML Schema required to provide for automated communication.

3 THE ONTOLOGICAL STRUCTURE OF MESSAGES IN MARITIME TRANSPORT

Ontology plays a major role in supporting the information exchange processes in maritime transport. In general, it provides a shared and common understanding of the domain of knowledge, communication, etc. The problem of ontology for maritime transport is mentioned, among others, in Malyankar et al. 1999, Mingyang P. et al. 2003, Kopacz et al. 2004 and Pietrzykowski et al. 2006.

One of the ways of message interchange is the use of radiotelephone VHF communication. The basic element of radiotelephone dialogs between objects such as ships, coast stations, etc., is a single message. Each message consists of at least one sentence. In practice sentences are usually simple ones and contain one piece of navigational information, e.g. for ship encounter situation (Fig. 2.):
- Alpha: 'Our CPA is close to 0',
- Alpha: 'Is it possible that we pass starboard to starboard?'

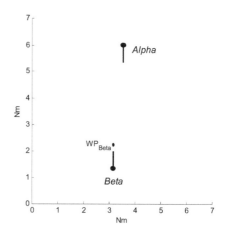

Figure 2. Ship encounter situation.

The sentences shown above contain one piece of navigational information – such piece is called an attribute in this context. Complex sentences - containing more than one attribute - may also be heard, e.g. two-attribute sentence: *Beta: „I intend to alter my course to starboard soon and cross ahead of you at a safe distance."* In this particular example one navigator informs of his intention to alter his ship's course to starboard and of the closest point of approach after the maneuver is completed.

In each sentence more than one attribute can be placed if they have the same simple sentence form when expressed separately. In other words, we cannot announce one piece of information in a sentence and ask about another piece of information in the same sentence.

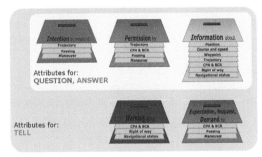

Figure 3. Sentence attributes divided into sentence forms.
Source: Pietrzykowski et al. 2006

Considering the forms of sentences, we should note that they significantly affect the meaning of formulated messages. A single message can be expressed as an interrogative and positive sentence. However, according to the accepted rules and using the recommendations concerning communication at sea (IMO 2002, IMO 2005) information is designed here as a group of attributes that can be linked to all possible types of sentences: Questions, Answers and Tells (statements).

Figure 3 shows that all information about intentions, permissions, information, warnings and requests can be expressed in the form of statements (Tell). The set of attributes related to intentions, permissions and information can be also provided in form of a Question (when we ask about e.g. permission for maneuver) or Answer (when we receive the permission for this maneuver).

The ontological structure of a message (Fig. 4) in the proposed automatic communication results from the structure of verbal communication and technical conditions:

– Header – supplemental data placed at the beginning of a block of data being transmitted, includes:
  – Sender – object sending a message (ship, coast station),

– Receiver – object(s) getting the message from sender (ships(s), coast station(s), objects located in circle- or square-shaped area),
– Sent – time of message casting,
– Other control information such as validity time, communication ID, information about message repetition.
– Body – information content of the message.

It is assumed that a message can be transmitted from the sender to a single destination (precisely defined address - unicast addressing), any group of interested destinations (multicast) or finally geocasted – to destinations identified by their geographical locations.

A) Message header

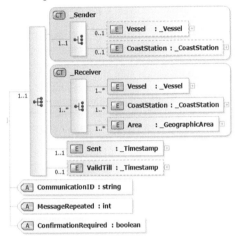

B) Message body

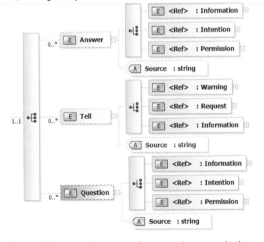

Figure 4. Structure of a message for automatic communication.

In the last case the location is pointed by rectangular or circular area described with geographical coordinates, where elevation is an optional parameter.

The body of the message consists of three groups of data related to all possible types of sentences: questions, answers and tells.

4 USE OF XML SCHEMA TO DESCRIBE ONTOLOGY

The process of developing the ontology and its result in the form of technical description of message syntax is a cyclic one. In each iteration of this model the following steps are required: updating requirements and analysis, design of ontology, implementation-testing of technical description of messages, maintenance. The result of the cycle is the richer version of both ontology and document description.

When the ontology for maritime transport communication is defined (the step of designing ontology is successfully made), it has to be described more precisely with constraints on the syntax and structure. It will allow generating and validating XML messages in an applied telecommunication system. XML Schema or DTDs can be used for that purpose. The XML Schema recommendation describes the content and structure of XML documents in XML. It includes the full capabilities of Document Type Definitions (DTDs), so that existing DTDs can be converted to XML Schema. Compared to DTDs, XML Schemas have additional capabilities.

According to the World Wide Web Consortium (W3C), XML Schema is itself represented in an XML vocabulary and uses namespaces, substantially reconstructs and considerably extends the capabilities found in XML document type definitions (DTDs).

XML Schema is itself represented in an XML vocabulary, whereas DTDs document is described in a unique syntax borrowed from SGML DTDs.

The size of message generated according to the description in XML Schema is about 50% larger than that based on DTDs. However, it does not seem to be a serious disadvantage, while its typical size is still several hundreds of characters and, if necessary, it can be compressed during transmission.

Finally, in some cases DTDs do not support the functionality required for XML documents, i.e. they do not ensure the compatibility with new XML products, do not support data types, and provide less complex constraints on the validity of XML documents. W3C recommend using XML Schema.

Therefore, in the following discussion it is assumed that XML Schema applies to the ontology in the way that allows automating building, validating and understanding of messages for communication in maritime transport.

```xml
<?xml version="1.0" encoding="utf-8"?>
<xs:schema elementFormDefault="qualified"
              xmlns:xs="http://www.w3.org/2001/XMLS
              chema">
  <xs:element name="Message">
   <xs:complexType>
    <xs:sequence>
     <xs:element name="Header">
      <xs:complexType>
        <xs:sequence minOccurs="1" maxOccurs="1">
        <xs:element name="Sender" type="_Sender"/>
        <xs:element name="Receiver" type="_Receiver">
        <xs:element name="Sent" type="_Timestamp" />
        <xs:element name="ValidTill" type="_Timestamp" />
        </xs:sequence>
        <xs:attribute            name="CommunicationID"
type="xs:string"
                              use="required" />
        <xs:attribute           name="MessageRepeated"
type="xs:int"/>
        <xs:attribute          name="ConfirmationRequired"
type="xs:boolean" />
      </xs:complexType>
     </xs:element>
     <xs:element         name="Body"         minOccurs="1"
maxOccurs="1">
      <xs:complexType>
       <xs:choice minOccurs="1" maxOccurs="1">
        <xs:sequence minOccurs="1" maxOccurs="1">
         <xs:element       name="Answer"       minOccurs="0"
maxOccurs="255">
          <xs:complexType>
           <xs:choice>
            <xs:element ref="Information" />
            <xs:element ref="Intention" />
            <xs:element ref="Permission" />
           </xs:choice>
           <xs:attribute name="Source">
            <xs:simpleType>
             <xs:restriction base="xs:string">
              <xs:enumeration value="Automatic" />
              <xs:enumeration value="Human" />
             </xs:restriction>
            </xs:simpleType>
           </xs:attribute>
          </xs:complexType>
         </xs:element>
[...]
</xs:schema>
```

Figure 5. Fragment of the ontology written in XML Schema.

Figure 5 shows the fragment of the more detailed ontology for maritime transport communication written in XML Schema. It is the technical form of message structure description for automatic communication that is shown in Figure 3. Its application in the real system allows to generate and validate messages.

A note is required about some XML Schema demands. All time stamps (type="_Timestamp") consist of combined date, time and time zone description. However, the time format is strictly required by

XML Schema definitions. It requires storing time value in form of *hh:mm:ss.ff* (*hh*-hours, *mm*-minutes, *ss*-seconds, *ff* –hundredths of a second).

We assume that some sentences can be fully-automatically exchanged between the sender and the receiver. Therefore, for each sentence additional information should be provided to indicate if a human or machine is the source of information. It is important when the communication is not only between system operators (humans) but is semi- or fully-automatic (between machines).

One of the results of the maritime transport ontology development is the message structure – a universal envelope that allows exchanging information among objects of the communication process (ships and all other types of watercraft, aircraft, coast stations, land vehicles). Although in the example mentioned in the next section the communication between two ships is described, more general communication can be processed (Fig. 6: note <Vessel> tags in both sender and receiver related lines in message headers).

5 APPLICATION

Let the dialog between two ships: *Alpha and Beta*, presented in the paper by Pietrzykowski at al. 2006, be an example – a case study – showing the communication described by XML messages generated according to the ontology described by XML Schema.

A situation. Both ships - *Alpha* and *Beta* - (Fig. 2) are in a situation that COLREGs qualify as "ships are on opposite courses or nearly opposite courses" – see Rymarz 1995. In this case, both ships are obliged to alter course to starboard (pass each other port-to-port) in order to safely pass each other. However, in certain conditions the regulations allow ships to alter their courses to port side, so that they pass each other on starboard sides.

Verbal communication. In our case, the ship *Alpha* suggested to the ship *Beta* that both ships pass on their starboard sides, as passing to port might have caused a dangerous situation due to the presence of another ship. In response, the *Alpha* received information that the *Beta* is about to alter course to starboard (because she approaches her waypoint), which will result in passing ahead of *Alpha* at a safe distance and the encounter situation will be solved. The *Alpha* accepts this solution.

Messages used in automated communication. The above dialog can be described in the form of XML messages built according to the ontology structure described in XML Schema (Figure 6).

The record also illustrates the membership of information kinds which are related to a given attribute. For example, attributes "Expectation", "Request" and "Demand" may be related to the same kind of information.

a)
```xml
<?xml version="1.0" encoding="utf-8"?>
<Message xmlns:xsi="http://www.w3.org/2001/XMLSchema-instance" (View source...)>
    <Header CommunicationID="AB02" MessageRepeated="0" ConfirmationRequired="0">
        <Sender><Vessel Name="Alpha" MMSI="231002300" /></Sender>
        <Receiver><Vessel Name="Beta" MMSI="262998700" /></Receiver>
        <Sent Date="2010-11-01" Time="18:00:01.00" Zone="UTC" />
    </Header>
    <Body>
        <Tell Source="Automatic">
            <Information>
                <Position Latitude="50'49,1'N" Longtitude="01'03,1'W" Altitude="0" />
            </Information>
        </Tell>
        <Tell Source="Human">
            <Warning>
                <CPA Value="0.1NM">Dangerous</CPA>
                <RightOfWay Whose="Indefinite" Who="We" Action="MustGiveWay" />
            </Warning>
        </Tell>
        <Question Source="Human">
            <Permision>
                <Passing Type="Opposite" Side="Starboard" Berth="0.5NM" />
            </Permision>
        </Question>
    </Body>
</Message>
```

b)
```xml
<?xml version="1.0" encoding="utf-8"?>
<Message xmlns:xsi="http://www.w3.org/2001/XMLSchema-instance" (View source...)>
    <Header CommunicationID="AB02" MessageRepeated="0" ConfirmationRequired="0">
        <Sender><Vessel Name="Beta" MMSI="262998700" /></Sender>
        <Receiver><Vessel Name="Alpha" MMSI="231002300" /></Receiver>
        <Sent Date="2010-11-01" Time="18:00:51.00" Zone="UTC" />
    </Header>
    <Body>
        <Answer Source="Human">
            <Intention>
                <Maneuver>
                    <AC Dir="Stbd" Value="40">
                        <Time Date="2010-11-01" Time="18:02:00.00" Zone="UTC" />
                    </AC>
                </Maneuver>
            </Intention>
        </Answer>
        <Question Source="Human">
            <Permission>
                <Passing Type="Cross" Side="Ahead"
```

```
Berth="1.2NM" />
    </Permission>
  </Question>
 </Body>
</Message>


c)  <?xml version="1.0" encoding="utf-8"?>
  <Message
  xmlns:xsi="http://www.w3.org/2001/XMLSchema-
  instance" (View source...)>
    <Header              CommunicationID="AB02"
  MessageRepeated="0" ConfirmationRequired="0">
      <Sender><Vessel              Name="Alpha"
  MMSI="231002300" /></Sender>
      <Receiver><Vessel            Name="Beta"
  MMSI="262998700" /></Receiver>
      <Sent   Date="2010-11-01"   Time="18:01:22.00"
  Zone="UTC" />
    </Header>
    <Body>
      <Answer Source="Human">
        <Permission>
          <Passing    Type="Cross"    Side="Ahead"
  Berth="1.2NM" />
        </Permision>
      </Answer>
    </Body>
  </Message>
```

Figure 6. A dialog between two ships written in XML – all
massages validated with the ontology described in XML
Schema.

Message a) is sent by the ship *Alpha*, and its body
consists of two positive sentences (position and
warning against a dangerous situation) and one interrogative sentence (permission for passing). Responding, the ship *Beta* sends message b), in which
she declares an intention of making a turning maneuver to starboard soon and asks for permission to
pass ahead of *Alpha* at a safe distance. The ship *Alpha* sends an answer (message c)) which includes
the permission for the proposed passing.

6 CONCLUSIONS

A general concept of communication system
for maritime transport was introduced. The cyclic
development process of ontology and its result –
technical description of message syntax (XML
Schema) – allow to build the solution in iterative
steps. Therefore, these authors proposed an ontological structure of messages and its description in
XML Schema to formalize format of contents of
messages. The general form of message envelope
was developed to support communication among
watercraft, aircraft, coast stations and land vehicles.
It is a basis for the development of the general ontology for maritime transport.

XML messages validated with partial maritime
ontology described in XML Schema were shown as
the example of implementation of communication
between two ships.

Further research will be focused on defining and
implementation of detailed ontology parallel to the
development of automatic negotiation system based
upon proposed automated communication system.

REFERENCES

Beam C., Segev A. 1997, Automated Negotiations: A Survey
of the State of the Art. In: *Wirtschaftsinformatik*: 263-268,
Vol. 39 (1997).
INTERNATIONAL MARITIME ORGANIZATION (IMO)
2002, Resolution A.918(22) adopted 29.11.2001, IMO Assembly 22nd Session 25.01.2002.
INTERNATIONAL MARITIME ORGANIZATION (IMO)
2005, Standard Marine Communication Phrases (English-
Polish edition), Maritime University of Szczecin, Szczecin.
Karp A. H. 2004, Rules of Engagement for Automated Negotiation. In: *proc. of the First IEEE International Workshop
on Electronic Contracting (WEC'04), San Diego, USA*: 32-
39.
Klein M., Fensel D., van Harmelen F., Horrocks I. 2001, The
relation between ontologies and XML schemas. In. *Linkö-
ping Electronic Articles in Computer and Information Sci-
ence*, Vol. 6, No. 004
Kopacz Z., Morgaś W., Urbański J. 2004, Information on
Maritime Navigation; Its Kinds, Components and Use. In:
European Journal of Navigation, vol. 2, no. 3, Aug 2004.
Malyankar R. M. 1999, Creating a Navigation Ontology.
In: *Tech. Rep. WS-99-13, AAAI Press*: 48-53, Menlo Park,
CA.
Mingyang P., Deqiang W., Shaopeng S., Depeng Z. 2003, Re-
search on Navigation Information Ontology. In: *proc. of the
11th IAN World Congress Smart Navigation – Systems and
Services*, October 2003.
Paurobally S., Turner P. J. and Jennings N. R. (2003), Auto-
mating negotiation for m-services. In: proc. of the *IEEE
Transactions on Systems, Man, and Cybernetics (Part
A: Systems and Humans), Special issue on M-services.
33(6)*: 709-724.
Pietrzykowski Z., Chomski J., Magaj J., Niemczyk G. 2006,
Exchange and Interpretation of Messages in Ships Commu-
nication and Cooperation System. In: Advanced in
Transport Systems Telematics, Ed. J. Mikulski, Publisher
Jacek Skalmierski Computer Studio, Katowice 2006, pp.
313-320.
Pietrzykowski Z., Magaj J., Chomski J. 2003, Sea-Going Ves-
sel Control in the Vessel Communications and Co-
Operation System. In: *Scientific Papers, Silesian University
of Technology, Transport No.51, Katowice 2003*, pp. 455-
462. Proc. of the 3rd International Conference Transport
Systems Telematics – 2003.
Pietrzykowski Z., Magaj J., Niemczyk G. 2002, Chomski J., A
sea-going vessel in an intelligent marine transport.
In: *Scientific Papers Silesian University of Technology,
Transport No. 45, Katowice 2002*, pp. 203-213. Proc. of the
2nd International Conference Transport Systems Telematics
– 2002.
Rymarz W., A handbook of Collision regulations (in Polish),
Published by Trademar, Gdynia 1995, Poland.

16. An Invariance of the Performance of Noise-Resistance of Spread Spetrum Signals

G. Cherneva & E. Dimkina
University of Transport, Sofia, Bulgaria

ABSTRACT: The paper is a study on the invariance of the performance of noise-resistance with respect to the quasi-determined spectrum-concentrated interferences with transmitting of spread spectrum signals with a fixed algorithm of processing.

1 INTRODUCTION

One of the main characteristics determining the effectiveness of a radio communication system is the stability against disturbances [1,2]. It is characterized with the dependency of the fidelity of received communications on the line energy parameters, algorithms used to transmit information and statistical characteristics of interferences [1,2]. With discrete systems of connections, the error probability of distinguishing signals is used for fidelity assessment [3,4].

The modern radio communication systems with safety responsibility are required to guarantee that the error-probability given will not exceed the preliminary specified permissible value independently of the variability activity of the channel. In essence it means that the given quantity of the system functioning is achieved thanks to the independency (partial or complete) of the performance of the noise-resistance from reasons causing the non-stationary state of the channel of connection. In the theory of automatic control this ability of the system to oppose resist against the disturbing actions is known as invariance. In the systems of connection, the part of disturbing effects is played by different disturbances and noise-resistance is the feature of the system that is invariant to them.

2 DEFINING THE CONDITIONS OF INVARIANCE

For a radio channel, the typical situation is the one where the performance of noise-resistance is determined by the presence of disturbances of several classes (fluctuating, spectrum-concentrated, impulse). The functional kind of the expression of the error probability with receiving by elements depends on the sets of signal parameters, the disturbances and the interaction between them:

$$P = \left[\left\{ \overline{h}_i^2 \right\} \left\{ \overline{h}_\zeta^2 \right\} \left\{ \overline{G}_{ij}^2 \right\} \right], \tag{1}$$

where $\overline{h}_i^2$ and $\overline{h}_\zeta^2$ express the mean statistical properties of the ratios between the energies of the *i-th* signal variant and the *j-th* disturbance variant and the white noise spectral density.

The part of parameters of interaction between the signal and disturbances is played by set $\{\overline{G}_{ij}^2\}$, $i=1÷n$, $j=1÷n_\zeta$, average statistical values of the coefficients of the reciprocal differences in the frequency-and-time area of their structures.

As the degree of the interaction between the useful signal and the disturbance on the frequency-and-time plane is analogous to their mutual correlation function, it is suitable to assume the average statistical value of the mutual difference coefficient in the position of interaction between them. This value is expressed in the kind of:

$$\overline{G}_{ij}^2 = \left[\frac{K_0 K_\zeta}{2P_i T} \int_0^T \dot{S}_i(t) \Sigma_{\zeta_j}^*(t) dt \right]^2, \tag{2}$$

where K_0, K_ζ, are the amplitude coefficients of the signal and disturbance, $T = \tau_0 N$ is the signal length, $\dot{S}_i(t)$ and $\Sigma_{\zeta_j}^*(t)$ are the complex functions of the *i-th* signal and *j-th* disturbance,

$P_i = \dfrac{K_0^2}{T}\displaystyle\int_0^T s_i^2(t)dt$ is the average power of the *i-th* signal variant.

The conditions of the invariance of the connection system are expressed in relation to a certain class of disturbances and in dependence with the metrics selected on the signal space.

If *n(t)* and $\zeta(t)$ are random realizations of fluctuation noise {N} and quasi-determined interferences {Ξ} respectively, then the performance of noise-resistance is a function of both interferences:

$$P = P(N, \Xi,). \tag{3}$$

The system of connection is absolutely invariant to $\zeta(t)$, if:

$$P(n, \zeta) = P(n, 0) = P(n) \tag{4}$$

is fulfilled.

When the noise-resistance characterization depends on interferences Ξ to a certain extent, e.g.:

$$|P(n, 0) - P(n, \zeta)| \le \varepsilon, \tag{5}$$

then the system is relatively invariant (invariant to ε), where ε presents the given distance between $P(n, \zeta)$ and $P(n, 0)$:

$$\varepsilon = \max_{\zeta} |P(n, \zeta) - P(n, 0)| \tag{6}$$

3 STUDY ON THE INVARIANCE OF THE PERFORMANCE OF NOISE-RESISTANCE IN REGARD TO SPECTRUM-CONCENTRATED INTERFERENCES WITH COHERENT RECEIVING SSS.

Under the condition of the effect only of fluctuating white noise {N} the noise-resistance of the system is determined by the ratio of signal energy power W1 to the spectrum density of white noise power v_0^2:

$$h^2 = \dfrac{W_1}{v_0^2}. \tag{7}$$

With transmitting opposed signals and fixed ratio signal/noise, the optimal operator of the receiver that is to ensure maximum noise-resistance against the interference {N} is the algorithm of coherent receiving with probability of error [3]:

$$P = \dfrac{1}{2}\left[1 - \Phi(h\sqrt{2})\right], \tag{8}$$

where $\Phi(.)$ is the integral function of distribution of Cramp.

With complicating the noise situation in the channel, when besides the fluctuating noise there are also effects caused by spectrum-concentrated interferences $\zeta(t)$, the probability of error is determined with independence [2]:

$$P = \dfrac{1}{2}\left[1 - \Phi(\sqrt{2}h_e)\right], \tag{9}$$

where h_e considerably depends on the type of receiver and the frequency- and time properties of the signals processed and the influencing disturbances. For a receiver optimal for channels with fluctuation noise and working with the influence of spectrum-concentrated disturbances:

$$h_e^2 = \dfrac{h^2}{1 + \bar{h}_\zeta^2 G_{ij}^2} \tag{10}$$

It is function $P(h)$ of the probability of error from parameter *h* that appears in the capacity of invariant of the system in relation to disturbance $\zeta(t)$. Taking into consideration dependencies (9) and (10), it is obtained for it:

$$P = \dfrac{1}{2}\left\{1 - \Phi\left[\sqrt{2}h\left(1 + \bar{h}_\zeta^2 G_{ij}^2\right)^{-\frac{1}{2}}\right]\right\}. \tag{11}$$

The characteristic of noise-resistance obtained depends of the effect of $\zeta(t)$ and $P = f\left(\bar{h}_\zeta^2\right)$. Therefore:

$$P(h, \zeta) \ne P(h, 0), \quad P(h) \ne in\,var\,\zeta(t),$$

i.e. the condition of absolute invariance (4) to spectrum-concentrated disturbances has not been satisfied.

The Table 1 gives the values of the probability of error calculated according to dependency (10) for the cases when there are no concentrated disturbances $(\bar{h}_\zeta^2 G_{ij}^2 = 0)$ and when a disturbance of $\bar{h}_\zeta^2 = 10^2$ и $G_{ij}^2 = 10^{-3}$ is influencing.

From the analysis of the data in Table 1 it follows that the maximum increase of the probability of error in area $P \le 10^{-2}$ is $6{,}259.10^{-3}$. According to the condition of invariance (3) it follows that the relative invariance feature of noise-resistance up to $\varepsilon = 6{,}259.10^{-3}$ is available in respect to spectrum-concentrated interferences .

When the concentrated interferences are of uniform spectrums, the coefficient of reciprocal difference G_{ij}^2 from the signal basis can be expressed as [2]:

$$G_{ij}^2 = \frac{\rho}{F_i T} \qquad i=1,2, \qquad (12)$$

where, with given $S_i(t)$ и $\Sigma_j(t)$, ρ is a constant quantity located in the interval:

$$1 \le \rho \le F_\zeta T \le F_i T ,$$

as the left limit of the interval is valid for a sinus-like shape of disturbance. With $F_i = F = const$, dependency (10) takes the kind of:

$$h_e^2 = \frac{h^2}{1 + \dfrac{\rho \overline{h}_\zeta^2}{FT}} \qquad (13)$$

Table 1. Values of the probability of error

$\overline{h}_\zeta^2 G_{ij}^2 = 0$	$\overline{h}_\zeta^2 G_{ij}^2 = 0.1$
$P(h)$	
0.5	0.5
0.079	0.089
0.023	0.028
$7.153 \cdot 10^{-3}$	$9.759 \cdot 10^{-3}$
$2.339 \cdot 10^{-3}$	$3.5 \cdot 10^{-3}$
$7.827 \cdot 10^{-4}$	$1.284 \cdot 10^{-3}$
$2.66 \cdot 10^{-4}$	$4.785 \cdot 10^{-4}$
$9.141 \cdot 10^{-5}$	$1.802 \cdot 10^{-4}$
$3.167 \cdot 10^{-5}$	$6.841 \cdot 10^{-5}$
$1.105 \cdot 10^{-5}$	$2.614 \cdot 10^{-5}$
$3.872 \cdot 10^{-6}$	$1.004 \cdot 10^{-5}$
$1.363 \cdot 10^{-6}$	$3.872 \cdot 10^{-6}$
$4.817 \cdot 10^{-7}$	$1.499 \cdot 10^{-6}$
$1.707 \cdot 10^{-7}$	$5.818 \cdot 10^{-7}$
$6.066 \cdot 10^{-8}$	$2.265 \cdot 10^{-7}$
$2.16 \cdot 10^{-8}$	$8.834 \cdot 10^{-8}$

It is in Fig. 1 where the dependency of the probability of error $P = f(h^2)$ determined by dependency (9) for different values of $\dfrac{\rho \overline{h}_\zeta^2}{FT}$ has been studied. Under the conditions of influence of only a fluctuating noise (Curve 1), the probability of error is determined only from h^2 regardless of the shape of signals that are transmitted. With $\dfrac{\rho \overline{h}_\zeta^2}{FT} \ge 10$ (curves 3 and 4), the efficiency of the coherent receiver under examination has been reducing considerably.

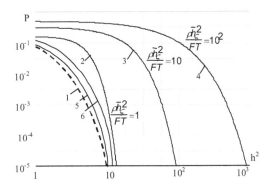

Fig.1. The dependency $P = f(h^2)$ for different cases

For the radio channels of decimeter range, the intensity of the concentrated interferences s is characterized by $\overline{h}_\zeta^2 = 10 - 10^4$. Hence, to provide invariance in respect to concentrated interferences and to guarantee the given noise-resistance level, it is necessary to use complex signals with a basis size depending on the ratio $\overline{h}_\zeta^2$.

The complicated noise background requires an optimization of the circuit of the receiver and adaptation of its structure depending on the interfering effects. In all known cases of systems designed with considering the effect of fluctuating noise and spectrum- concentrated interferences [2], the expressions of the probability of error depend monotonously on the value of the product for random j:

$$\delta_j = \frac{G_{0j}^2 \overline{h}_{\zeta j}^2}{1 + \overline{h}_{\zeta j}^2} , \qquad (14)$$

so that

$$h_e^2 = h^2(1 - \delta_j) . \qquad (15)$$

Taking into account (14) and (15), the probability of error can be expressed as:

$$P = \frac{1}{2}\left\{1 - \Phi\left[\sqrt{2}h\left(1 - \frac{\overline{h}_\zeta^2 G_{0j}^2}{1 + \overline{h}_\zeta^2}\right)^{\frac{1}{2}}\right]\right\} \qquad (16)$$

With an intensity of the spectrum-concentrated interferences , such as $\overline{h}_\zeta^2 \ge 10$, it is obtained that :

$$\frac{h_\zeta^2}{\left(1 + h_\zeta^2\right)} \approx 1$$

and $\delta_j \leq G_{0j}^2 = const\left(h_\zeta^2\right)$.

Hence h_e^2 does not depend on $\overline{h}_\zeta^2$ and the optimized receiver is invariant with regard to the influencing spectrum-concentrated interferences   unlike the receiver of nature (8). Besides that, for signals of sufficiently big bases $\delta_j \leq \dfrac{\rho}{FT} \ll 1$ and $h_e^2 \approx h^2$ has been provided, i.e. in practice, the noise-resistance of that receiver does not differ from the noise-resistance in channels only of fluctuating noise. In Fig.1 what has been studied is the dependency of the probability of error from h^2 with two values of coefficient δ_j - curve 5 (with $\delta_j = \dfrac{1}{4}$) and curve 6 (with $\delta_j = \dfrac{1}{10}$).

4 CONCLUSIONS

The paper presents the obtained analytical dependencies between the size of the basis of the transmitted complex signals and the possibilities of coherent demodulators to compensate the effect of interferences. A coherent demodulator optimal in respect to white noise and an optimized demodulator operating under the conditions of the effect of white noise spectrum-concentrated   interferences   have   been compared. The results obtained have shown that the optimized receiver can keep a fixed level of probability of error with considerably smaller signal bases providing an absolute invariance of the nature of noise-resistance against spectrum-concentrated interferences.

REFERENCES

1. Proakis, J. 2001. Digital Communications. Mc Graw Hill Series in El.Eng. Stephen W.
2. Haykin, S. 1994.Communication Systems. Willey & Sons, USA.
3. Simon, M.K., J.K. Omura, R.A. Sholtz. 2001. Spread Spectrum Communications. Handbook, Hardcover.
4. Holmes, J.K. 1982. Coherent Spread Spectrum Systems. Willey, New York.

17. Surface Reflection and Local Environmental Effects in Maritime and other Mobile Satellite Communications

S. D. Ilcev

Durban University of Technology (DUT), Durban, South Africa

ABSTRACT: This paper introduces the effects of surface reflections and local environmental as very important particulars for mobile and especially for Maritime Satellite Communications (MSC), because such factors generally tend to impair the performance of satellite communications links, although signal enhancements are also occasionally observed. Local environmental effects include shadowing and blockage from objects and vegetation near the Ship Earth Station (SES) and other mobile terminals. The advantages and disadvantages of those effects are discussed, the areas of surface reflection were examined and the further investigations of local environmental are provided. At this point, surface reflections are generated either in the immediate vicinity of the SES terminals or from distant reflectors, such as mountains and large industrial infrastructures. Specific issues related to these challenges are concluded and a set of solutions is proposed to maximize the availability of satellite communication capacity to the mobile user applications. The specific effects on propagation in the mobile environments are examined and explained important characteristics of the Interference from Adjacent Satellite Systems, Specific Local Environmental Influence in MSC, different Noise Contribution of Local Ships' Environment, Blockages Caused by Ship Superstructures and Motion of Ship's Antenna.

1 INTRODUCTION

The reflected transmission signal can interfere with the direct signal from the satellite to produce unacceptable levels of signal degradation. In addition to fading, signal degradations can include intersymbol interference, arising from delayed replicas. The impact of the impairments depends on the specific application, namely in the case of typical Land MSC (LMSC) links, all measurements and theoretical analysis indicate that the specular reflection component is usually negligible for path elevation angles above 20°.

Moreover, for handheld terminals, specular reflections may be important as the low antenna directivity increases the potential for significant specular reflection effects. For Maritime MSC (MMSC), LMSC and Aeronautical MSC (AMSC) system links, design reflection multipath fading, in combination with possible shadowing and blockage of the direct signal from the satellite, is generally the dominant system impairment.

2 REFLECTION FROM THE EARTH'S SURFACE

Prediction of the propagation impairments caused by reflections from the Earth's surface and from different objects (buildings, hills, vegetation) on the surface is difficult because the possible impairment scenarios are quite numerous, complex and often cannot be easily quantified. For example, the degree of shadowing in LMSC satellite links frequently cannot be precisely specified.

Therefore, impairment prediction models for some complicated situations, especially for LMSC links, tend to be primarily empirical, while more analytical models, such as those used to predict sea reflection fading, have restricted regions of applicability. Nevertheless, the basic features of surface reflections and the resultant effects on propagating signals can be understood in terms of the general theory of surface reflections, as summarized in the following classification:

1 Specular Reflection from a Plane Earth – Here, this effect is less than or equal to the coefficient for horizontal polarization. Thus the polarization of the reflected waves will be different from the polarization of the incident wave if the incident polarization is not purely horizontal or purely ver-

tical. For example, a circularly polarized incident wave becomes elliptically polarized after reflection.

2 Specular Reflection from a Smooth Spherical Earth – Here, the incident grazing angle is equal to the angle of reflection. The amplitude of the reflected signal is equal to the amplitude of the incident signal multiplied by the modules of the reflection coefficient.

3 Divergence Factor – When rays are secularly reflected from a spherical surface, there is an effective reduction in the reflection coefficient, which is a geometrical effect arising from the divergence of the rays.

4 Reflection from Rough Surface – In many practical cases, the surface of the Earth is not smooth. Namely, when the surface is rough, the reflected signal has two components: one is a specular component, which is coherent with the incident signal, while the other is a diffuse component, which fluctuates in amplitude and phase with a Rayleigh distribution.

5 Total Reflected Field – The total field above a reflecting surface is a result of the direct field, the coherent specular component and the random diffuse component.

6 Reflection Multipath – Owing to the existence of surface reflection phenomena signals may arrive at a receiver from multiple apparent sources. Thus, the combination of the direct signal (line-of-sight) with specular and diffusely reflected waves causes signal fading at the receiver. The resultant multipath fading, in combination with varying levels of shadowing and blockage of the line-of-sight components, can cause the received signal power to fade severely and rapidly for MES and is really the dominant impairment in the Global Mobile Satellite Communications (GMSC) service.

3 FADING IN MMSC AND AMSC SYSTEMS DUE TO SEA SURFACE REFLECTION

Multipath fading due to sea reflection is caused by interference between direct and reflected radiowaves. The reflected radiowaves are composed of coherent and incoherent components, namely specular and diffuse reflections, respectively, that fluctuate with time due to the motion of sea waves. The coherent component is predominant under calm sea conditions and at low elevation angles, whereas the incoherent becomes significant in rough sea conditions. If the intensity of the coherent component and the variance of the incoherent component are both known, the cumulative time distribution of the signal intensity can be determined by statistical consideration.

In any event, a prediction model for multipath fading due to sea reflection, however, was first developed for MMSC systems at a frequency near 1.5 GHz. Although the mechanism of sea reflection is common for MMSC and AMSC systems, only with the difference that fading characteristics for AMSC are expected to differ from those for MMSC, this is because the speed and altitude of aircraft so much greater than those of ships. At this point, the effects of refractions and scattering by the sea surface become quite severe in case of MMSC and AMSC, particularly where antennas with wide beam widths are used.

The most common parameter used to describe sea condition is the significant wave height (H), defined as the average value of the peak-to-trough heights of the highest one-third of all waves. Empirically, H is related to the r.m.s. height (h_o) by:

$$H = 4h_0$$

Hence, at 1.5 GHz the smaller-scale waves can be neglected and the r.m.s. value of the sea surface slopes appear to fall between 0.04 to 0.07 in the case of wave heights less than 4 m.

Thus, with diminishing satellite elevation angle, the propagation path increases, causing a decrease of signal power at the Rx side. The noise level is initially constant; however, upon reaching some critical value of the elevation angle, sea-reflection signals appear at the Rx input, which begins to affect the C/N value. To include the effect of multipath interference caused by sea-refracted signals, the reception quality would be more properly described by the C/N plus M, where M is an interfering sea-reflected signal acting as a disturbance. Thus, sea-reflected signals differ in structure and can be divided into two categories:

1 Radio signals with the rapid continuous fluctuations of amplitudes and phases and with a possible frequency shift due to the motion of small portions of the specular cross-section relative to the source of signals (noise or diffused components).

2 Radiowaves with relatively slowly changing phase close to the phase of the basic signal and with an amplitude correlating with that of the basic signal (specular component).

Consequently, within the overall specular cross-section, an angle of arrival reflected radio signals relative to the horizontal plane may be regarded as constant and can be described by the following expression:

$$\alpha = 90° - \gamma$$

where α = angle of radio signals arrival in accordance with Figure 1. and γ = reflection angle. The modulus of sea reflection factor for L-band signals is within 0.8 and 0.9, which means that the amplitude of the specular reflected signal is nearly the same as that of the direct signal. As measurements have shown, the noise component depends only upon an elevation angle and a wave height. Decreasing the elevation angle and increasing the wave height result in an increase in the total amplitude of the noise, which includes the noise component. At elevation angles below 5°, the amplitude component reaches a peak value and is no longer affected by the wave height. Now an increase of the wave height causes primarily more frequent variations in the noise component. The corresponding deviation of C.N measured in 1 kHz bandwidth amounts from 4.5 to 5 dB.

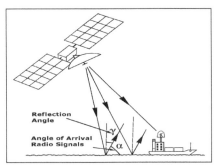

Figure 1. Geometry of Sea Reflection of Satellite Radio Signals
Courtesy of Book: "Global Mobile Satellite Communications" by D.S. Ilcev

The specular component that appears at the Rx input together with the direct signal causes fading in the direct signal due to both the minor difference between their phases and the slow change of the parameters of the reflected signals. The ratio of the direct to specular reflected signal can be described as:

$$C/M = (C + G_\varepsilon) - [C - G_{(\alpha + \varepsilon)}]$$

where C = direct signal; M = specular reflected signal power from the sea; G_ε = maximum gain of the receive SES antenna pointing towards the satellite; ε = elevation angle and α, as is shown in Figure 1. In addition, keeping accuracy sufficient for practical purposes, the previous relation gives:

$$C/M = \beta_{C/N} + [G_\varepsilon - G_{(\alpha + \varepsilon)}]$$

where $\beta_{C/N}$ = deviation of C/N ratio. With decreasing elevation angle, the C/M diminishes monotonically, except for the elevation angle range of 5° to 8°, within which a rise in C/N is observed. This is obviously due to the fact that at the said angles the difference in path between the direct and the specular signal becomes negligible, so that conditions appears close to the summation of the similar signals at the receiver input. An increase of the C/N plus M ratio is observed simultaneously due to reaching a peak value of amplitude in the noise (diffused) component. In fact, experimental measurements show that as the elevation angle decreases from 10° to 1°, the mean C/N plus M diminishes from 22–24 dB to 17–18 dB,

with the deviation increasing from 1.5–2 dB to 4.5–5 dB.

4 MULTIPATH FADING CALCULATION MODEL FOR REFLECTION FROM THE SEA

The amplitude of the resultant signal at the SES terminal, being the sum of the direct wave component, the coherent and the incoherent reflection components, has a Nakagami-Rice distribution (see ITU R P.1057). The cumulative distribution of fading depends on the coherent-to-incoherent signal intensities. For example, in the case of rough sea conditions at 1.5 GHz, the coherent reflection from the sea is virtually non-existent and the coherent signal is composed only of the direct component. Therefore, the fading is determined by the Carrier-to-Multipath ratio (C/M), i.e., the power ratio of the direct signal and multipath component caused by incoherent reflection. The maximum fade depth (Φ_{max}) occurs when the coherent multipath signal is in anti-phase with the direct signal, given by:

$$\Phi_{max} = - 20 \log (1 - A_r) \quad [dB]$$

where A_r = amplitude of the coherently reflected component. The value decreases rapidly with increasing wave height, elevation angle and RF. In practice, due to the vertical motion of the ship antenna relative to average sea surface height, the maximum fade value will seldom occur. By adding Φ_{max} and $\Phi_i(p)$ as signal fade due to the incoherent component in the function of time percentage (p), a practical estimate of the combined fading effects of the coherent and incoherent multipath signal for sea conditions is obtained:

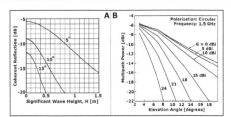

Figure 2. Estimates of Coherent Reflection and Multipath Power
Courtesy of Book: "Mobile Antenna Systems Handbook" by K. Fujimoto and J.R. James

$$\Phi_c = \Phi_{max} + \Phi_i(p)$$

The maximum fade value due to the coherent component will not occur constantly because of the vertical motion of the ship antenna relative to average sea surface height; therefore, the estimate using this equation seems to give the worst-case value. In practice, for low elevation angles (les than 10°) at around L-band frequencies, the maximum fading occurs when the significant sea wave height is between 1.5 and 3 m, where the coherent reflected component is negligible. Accordingly, the dependence of fading depth on wave height in this range is relatively small.

The amplitude level of the coherent component decreases rapidly with increasing sea wave height, elevation angle and frequency. Figure 2. (A) illustrates the relationship between coherent reflection and significant wave height. Namely, estimates of amplitude of the coherent component for an omnidirectional antenna as a function of a significant wave height for low elevation angles are illustrated; the frequency is 1.5 GHz and polarization is circular. Thus, the incoherent component is random in both amplitude and phase, since it originates from a large number of reflecting facets on the sea's waves. The amplitude of this component follows Rayleigh distribution and the phase has a uniform distribution.

Since the theoretical model concerning the incoherent components is not suitable for engineering computations using a small calculator, simpler prediction models are useful for the approximate calculation of fading. Such simple methods for predicting multipath power or fading depth have been recently developed. Thus, in Figure 2. (B) is presented the relationship between multipath power and elevation angle for different antenna gains. Although fading depth depends slightly on sea surface conditions, even if the incoherent is dominant, the simple model is useful for a rough estimate of fading depth.

Fading depth, which is a scale of intensity of fading, is usually defined by the difference in decibels between the direct wave signal level and the signal level for 99% of the time. The fading depth can be approximated by a 50% to 99% value for fading where the incoherent component is fully developed. Large fading depths usually appear in rough sea conditions, where the incoherent component is dominant. Thus, Figure 3. (A) shows the fading depth estimated by the simple method for antenna not exceeding for 99% of the time and the corresponding C/M ratio for circular polarization at 1.5 GHz band under the condition of significant wave heights from 1.5 to 3 m. The antenna gains of 24, 20, 15 and 8 dB are functions of elevation angle with a fully-developed incoherent component. The calculation is based on the theoretical method, where the shaded area covers the practical range of the sea wave slope, which depends on fading depth in rough sea conditions. Values estimated by this simple method give the mean values of those given in Figure 3. (B).

On the other hand, as the theoretical model is not suitable for engineering computations using a small calculator, these simple prediction models are really useful for the approximate calculation of fading or interference. Such simple methods for predicting multipath power or fading depth have been developed by Sandrin and Fang [1986] and by Karasawa and Shiokawa [1988] for MMSS and Karasawa [1990] for AMSS.

Furthermore, the frequency spectral bandwidth of temporal amplitude variations enlarges with increasing wave height and elevation angle. Figure 3. (B) shows the probable range of −10 dB spectral bandwidth (which is defined by the frequency corresponding to the spectral power density of −10 dB relative to the flat portion of power spectrum) of L-band multipath fading obtained by the theoretical fading model as a function of the elevation angle under the usual conditions of MMSC; namely, significant wave height of 1 m to 5 m, ship speed of 0 to 20 knots and rolling conditions of 0 to 30°.

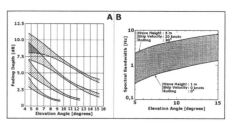

Figure 3. Estimates of Fading Depth and Spectral Bandwidth Courtesy of Book: "Mobile Antenna Systems Handbook" by K. Fujimoto and J.R. James

5 OTHER ESTIMATIONS OF FADING FOR MMSC AND AMSC SYSTEMS

The error pattern in digital transmission systems affected by multipath fading is usually of the burst type. Accordingly, a firm understanding of the fade duration statistics of burst type fading is required. Mean value of fade duration (Φ_D) and fade occurrence interval (Φ_o) for a given threshold level as a function of time percentage, can be estimated from the fading spectrum. A simple method for predicting the mean value from the −10 dB spectral bandwidth is available as a theoretical fading model.

Predicted values of (Φ_D) and (Φ_o) for 99% of the time at an elevation angle from 5 to 10° are 0.05 to 0.4 sec for (Φ_D) and 5 to 40 sec for (Φ_o). The probability density function of (Φ_D) and (Φ_o) at any percentages ranging from 50% to 99% approximates an exponential distribution.

1 Simple Prediction Method of Fading Depth – According to theoretical analysis and experimental results made by the mentioned researchers in Japan, the lowest elevation angle Earth-to-space path at 1.5 GHz RF band satisfies the energy conservation law: [Power of coherent component] + [Average power of incoherent component] ~ Constant. If this expression is satisfied, the maximum incoherent power can be estimated easily by calculating the coherent power at u = 0. Otherwise, for a more accurate estimation, small modifications of some parameter dependencies

are necessary. The modified procedure has been adopted in P.680 for MMSC and P.682 ITU-R Recommendations for AMSC. Figure 4. (A) shows a scattergram of measured and predicted fading depths (i.e., fade for 99% of the time relative to that for 50%) in the case of MMSC systems between measured data and predicted values derived from the simple calculation method with the same conditions. In this figure, Φ_{dp2} are values from the method set out in ITU-R P.680, while Φ_{dp1} are those from an alternative procedure of the prediction method for scattering angles. It is evident that the values given by these methods agree well with the experimental values although the methods are rather approximate.

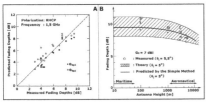

Figure 4. Scattergram and Altitude Dependence of Fading Depth Courtesy of Handbook: "Radiowave Propagation Information for Predictions for Earth-to-Space Path Communications" by ITU

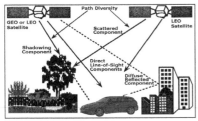

Figure 5. Typical LMSC Propagation Environment Courtesy of Book: "Mobile Satellite Communications" by S. Ohmori and other

In Figure 4. (B) is shown the altitude dependence of signal fade depth not exceeded for 99% of the time vs. antenna height on board ships or aircraft. This experiment was obtained from measurements with a helicopter together with the calculated values from the simple estimation method of the solid line and the theoretical model of the shaded region in the figure. From the figure it can be seen that the simple prediction method agrees well with both the theoretical model and measured data even in the case of the AMSC system.

2 Fading Spectrum – In system design, particularly for digital transmission systems, it is important not only to estimate the fading depth but also to know the properties of temporal variation, such as the frequency power spectrum. For MMSC systems, theoretical analyses were carried out in Japan and all parameters affecting the spectrum

such as wave height, wave direction, ship's direction and velocity, path elevation angle and antenna height variations due to ship's motion (rolling and pitching) were taken into account. In general, spectrum bandwidth is broader with increasing wave height, elevation angle, ship velocity and the relative motion of the ship borne antenna. The dependence of the spectral shape on antenna polarization and gain is usually very small. Moreover, since the speed of aircraft is significantly higher than that of ships, the fluctuation speed of multipath fading in AMSC is much faster than that in MMSC, depending on the flight elevation angle measured from the horizontal plane. The calculated –10 dB spectral bandwidth is between 20 and 200 Hz for elevation angles of 5° to 20°, for flight elevation angles 0° to 5° at a speed of 1,000 km/h.

6 FADING IN LMSC SYSTEM DUE TO SIGNAL BLOCKAGE AND SHADOWING

Recently, in the USA, Canada, Australia, Japan and Europe, domestic LMSC and GMPSC services have started. The main purpose of these systems is to extend MSC voice services to rural/remote areas where terrestrial/cellular services are not provided. In a typical urban environment, line-of-sight for cellular systems sometimes is not available due to blockage by buildings and other structures, when a MES can receive many waves reflected from these structures and conduct communication link using these signals.

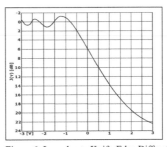

Figure 6. Loss due to Knife-Edge Diffraction Courtesy of Handbook: "Radiowave Propagation Information for Predictions for Earth-to-Space Path Communications" by ITU

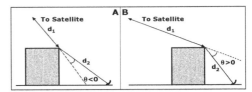

Figure 7. Geometry of Knife-edge Diffraction Phenomena Courtesy of Handbook: "Radiowave Propagation Information for Predictions for Earth-to-Space Path Communications" by ITU

The LMSC system can be expected to use the direct line-of-sight signal from a satellite because of the high elevation angles. When the line-of-sight is blocked by any obstacle, MSC is not available, but using path diversity the link is available because of overlapping of two or more signal from adjacent satellites. At present, path diversity from separate satellites is rarely used for GEO MSS, but Non-GEO MSS have an inherent capability to exploit diversity, because the number of satellites is large providing path diversity. At this point, Figure 5. presents a typical propagation environment for scattering, multipath fading, shadowing, diffusion, ect.

Therefore, to design LMSC system, one needs information about the propagation statistics of multipath fading and shadowing. A vehicle runs at a distance of 5 to 20 m from roadside obstacles using an omnidirectional antenna, which has azimuthally uniform gain but elevation directivity, or a medium or high-gain antenna with automatic tracking capability. Thus, signal blockage and shadowing effects occur when an obstacle, such as roadside trees, overpasses, bridges, tunnels, utility poles, high buildings, hills or mountains, impedes visibility to the focus of satellite. This results in the attenuation of the received signal to such an extent that transmissions meeting a certain quality of service may not be possible. At any rate, in the shadowing environments the presence of the trees will result in the random attenuation of the strength of the direct path signal. Hence, the depth of the fade is dependent on a number of parameters including tree type, height, as well as season due to the leaf density on the trees. Whether a VES is transmitting on the left or right-hand side of the road could also have a bearing on the depth of the fade, due to the line-of-sight path length variation through the tree canopy being different for each side of the road. In fact, fades of up to 20 dB at the L-band may be presented due to shadowing caused by roadside trees. This shadowing by roadside trees cannot occur on modern highways because they are free of trees, only sometimes can shadowing appear by tunnels, very big constructions, bridges and mountains or hills in narrow passages.

1 **Tree Shadowing** – Attenuation due to trees nearby the roads arises from absorption by leaves and blockage by trunks and branches. Absorption by leaves is a function of the type and size of leaves and the water content therein. Blockage due to trunks is primarily a function of their size. In addition to attenuation of the direct signal, trees also cause an incoherent component due to signals reflected and diffracted off the tree surfaces.

The overall attenuation from different types of fully foliated trees varies from 10.6 to 14.3 dB and the attenuation coefficient is from 1.3 to 1.8 dB/m. Measurements were conducted with MES Rx in a rural environment. Based on these average values, a frequency scaling law for the attenuation coefficient has been derived [Goldhirsh and Vogel] by:

$$a_1 = a_0 \sqrt{f_1/f_0} \quad [dB/m]$$

where a_1 and a_0 = attenuation coefficients at frequencies f_1 and f_0 [GHz], respectively. Hence, the range of variation for a_1 at 1.5 GHz is from 0.5 to 1.7 dB/m. Moreover, trees without foliage attenuate less and the reduction in attenuation appears to be proportional to the total attenuation experienced when the tree is fully foliated. The received strength of the direct signal behind a tree will depend on the orientation of the signal path with respect to the tree. The amount of absorbing matter lying along the path will determine the degree of attenuation and hence, on average, the length of the signal path through the tree shield can be considered a major factor in determining the signal level. The path length is a function of the elevation angle and the distance between the receiver and the tree. At this point, the average attenuation behind an isolated tree can be estimated as the product of the attenuation coefficient and the path length through the tree. The path length through the tree canopy will depend on its shape and the orientation of the signal path within the canopy. Depending on the type being considered, the tree canopy may be modeled as any one of the shapes.

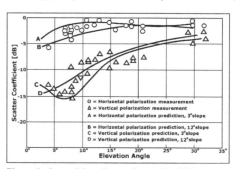

Figure 8. Ocean Mean-Square Scatter Coefficients vs. Elevation
Courtesy of Handbook: "Radiowave Propagation Information for Predictions for Earth-to-Space Path Communications" by ITU

For the intermediate elevation angle (20° to 50°), attenuation is almost independent of elevation and dependence becomes important only at the higher and lower ends of the elevation angle range. By considering the path length variability as a statistical parameter, however, a tree can be modeled as giving an average attenuation and a distribution around it. Both the coherent and incoherent components will vary with the receiver position and complete decorrelation of the signal is expected over distances in the order of a few wavelengths.

2 Building Shadowing – Signal reception behind buildings takes place mainly through diffraction and reflection. A direct line-of-sight component does not usually exist and therefore shadowing cannot be defined unambiguously, as in the case of trees. However, shadowing may be loosely defined as the power ratio between the average signal levels to the unscheduled direct signal level. Otherwise, diffractions from buildings can be studied using knife-edge diffraction theory, which gives reasonable estimates. A concept view of knife-edge diffraction phenomena is shown in Figure 6. for all losses caused by the presence of the obstacles as a function of a dimensionless parameter ν and in Figure 7., which illustrates the geometry of the path for both the illuminated (A) and shadowed cases (B), in order to calculate the parameter ν, by using elevation and wavelength as follows:

$$\nu = \varepsilon \sqrt{2} / [\lambda(1/d_1 + 1/d_2)] \quad \text{but } d_1 \gg d_2, \text{so: } \nu = \varepsilon \sqrt{2}/\lambda \ (1/d_2) = \varepsilon \sqrt{2d_2}/\lambda$$

The signal strength at the shadow boundary is 6 dB below the line-of-sight level. In the illuminated region, the signal fluctuations are experienced due to interference between the direct and the diffracted components. Hence, once inside the shadowed region, the shadow increases rapidly. An experimental investigation into building shadowing loss conducted by Yoshikawa and Kagohara in 1989 confirmed the applicability of the knife-edge diffraction theory. Measured signal strength behind a building at various distances was found to follow the prediction made, assuming a single diffraction edge. However, where the building is narrow compared with its height, there may be significantly less shadowing than predicted by the above procedure. When the direct signal path is blocked by a building, diffractions of the buildings are not expected to play a dominant role in establishing the communication link, unless MES is close to the shadow boundary. Reflections may play a useful role in such situations, as happens in cellular systems. Building penetration depends on the type of exterior material of the building and the location inside the building. Thus, the loss through the outer structure, known as the penetration loss, is defined as the difference in median signal levels between that measured immediately outside the building at 1.5 m above the ground and that immediately inside the buildings at some reference level on the floor of interest.
Measurements made at 940 MHz in a medium-size city in the USA indicate that on the ground floor of typical steel-concrete-stone office buildings, the average penetration is about 10 dB with a standard deviation of about 7 dB. While another set of measurements in a large city resulted in average ground floor penetration loss of 18 dB with

a standard deviation of 7.7 dB. However, the overall decrease of penetration loss with height was about 1.9 dB per floor. The biggest average attenuations are 12 dB and 7 dB and standard deviations are 4 db and 1 dB for metal and concrete, respectively.

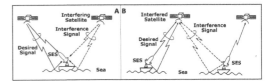

Figure 9. Basic Model for Intersatellite Interferences Phenomena
Courtesy of Handbook: "Radiowave Propagation Information for Predictions for Earth-to-Space Path Communications" by ITU

Attenuation through glass ranges from about 2 to 6 dB depending on the type of glass, i.e., plain glass produces less attenuation compared to tinted or coated glass, containing metallic components. Otherwise, the smallest average attenuation is through office furnishings, aluminum and wood/brick of about 1, 2 and 3 dB, respectively. Losses within a building are both of distance from the exterior wall blocking the signal path, as well as the interior layout. Measurements have resulted in an inverse distance power law coefficient ranging from 2 to 4. The forthcoming ICO system has conducted experiments with satellite-borne signals whose final target is to improve and even to eliminate building shadowing.

7 FADING IN AMSC SYSTEM DUE TO LAND REFLECTION

An experiment aboard a helicopter over land was carried out by receiving right-hand circularly polarized 1.5 GHz beacon signals from an IOR Marisat satellite at an elevation angle of 10°. Fading depths measured over plains such as paddy fields were fairly large (about 5 dB), nearly equal to that for sea reflection. However, fade depths measured over mountainous and urban areas were less than 2 dB. In the case of mountains, reflected waves are more likely to be shadowed or diffused by the mountains. As to urban areas, the shadowing and diffusing effects of reflected wave by buildings are also large. For this reason, the ground reflected multipath fading in these cases is not generally significant.
Measurements of Sea-Reflection Multipath Effect – A study of multipath propagation at 1.5 GHz was performed with KC-135 aircraft and the NASA ATS-6 satellite. Otherwise, the signal characteristics were measured with a two-element waveguide array in the aircraft noise radome, with 1 dB beam width of 20° in azimuth and 50° in elevation. Namely, data

was collected over the ocean and over land at a normal aircraft altitude of 9.1 km and with a nominal speed of 740 km/h. Coefficients for horizontal and vertical antenna polarization were measured in an ATS-6 experiment, where values for r.m.s. sea surface slopes of 3o and 12o were plotted versus elevation angle, in Figure 8., along with predictions derived from a physical optics model. Sea slope was found to have a minor effect for elevation angles above about 10°. The agreement between measured coefficients and those predicted for a smooth flat Earth as modified by the spherical Earth divergence factor increased as sea slope decreased.

The relationship between r.m.s. sea surface and wave height is complex but conversion can be performed. Namely, for most aeronautical systems, circular polarization will be of greater interest than linear. For the simplified case of reflection from a smooth Earth, which should be a good assumption for elevation angles above 10°, circular co-polar and cross-polar scatter coefficients (S_c and S_x), respectively, can be expressed in terms of the horizontal and vertical coefficients (S_h and S_v), respectively by:

$$S_c = (S_h + S_v)/2 \quad \text{and} \quad S_x + (S_h - S_v)/2$$

For either incident RHCP or LHCP is employed. Thus in general, the horizontal and vertical coefficients are complex values and phase information is required to apply the last equitations to the curve, in Figure 8.

8 INTERFERENCE FROM ADJACENT SATELLITE SYSTEMS

In GMSC systems for ships, vehicles and aircraft, small mobile antennas are essential for operational and economic reasons. As a result, a number of low G/T value MES terminals with smaller antennas have been developed. However, such antenna systems are subject to the restriction of frequency utilization efficiency, or coexistence between two or more satellite systems in the same frequency band and/or an overlap area where both satellites are visible.

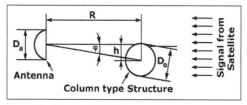

Figure 10. Geometry of Blocking
Courtesy of Handbook: "Radiowave Propagation Information for Predictions for Earth-to-Space Path Communications" by ITU

For coordination between two different systems in the same frequency band, a highly reliable interference evaluation model covering both interfering and interfered with conditions is required. Investigation into this area has been undertaken in particular by ITU-R Study Group 8. Advancement of such a model is an urgent matter for the ITU-R considering the number of MSC systems that are being developed in the meantime.

In GMSC systems, the desired signal from the satellite and the interfering signal from an adjacent satellite independently experience amplitude fluctuations due to multipath fading, necessitating a different treatment from that for fixed satellite systems. The main technical requirement is a formulation for the statistics of differential fading, which is the difference between the amplitude of the two satellite signals. At this point, the method given in No 5 of ITU-R P.680 Recommendation therefore presents a practical prediction method for signal-to-interference ratio where the effect of thermal noise and noise-like interference is taken into account; assuming that the amplitudes of both the desired and interference signal affected by the sea reflected multipath fading follow Nakagami–Rice distributions. In fact, this situation is quite probable in MMSC systems.

The basic assumptions of the intersatellite model are shown in Figure 9., as an example of interference between adjacent satellite systems, where (A) is downlink interference on the MES side and (B) is uplink interference on the satellite side. This applies to multiple systems sharing the same frequency band. It is anticipated that the interference causes an especially severe problem when the interfering satellite is at a low elevation angle viewed from the ship presented in this figure because the maximum level of interference signal suffered from multipath fading increases with decreasing elevation angle. Another situation is interference between beams in multispot-beam operation, where the same frequency is repeatedly allocated.

9 SPECIFIC LOCAL ENVIRONMENTAL INFLUENCE IN MMSC

Local environmental influence is important for SES equipped with beam width antenna. Many factors, with different kinds of noise sources, tend to make disturbances in MMSC channels. Another factor that affects communication links is RF emission from different noise sources in the local environment. Specific local ship environmental factors can be noise contributions from various sources in the vicinity of the SES and the influence of the ship's superstructure in the operation of maritime mobile terminals. However, some of these local environmental factors can affect SES when a ship is passing nearby the coast and, some of these are permanent noise sources. More exactly, these environmental sources include broadband noise sources, such as electrical equipment and motor vehicles and out-of-band emission from powerful transmitters such as radars and ships HF transmitters.

9.1 *Noise Contribution of Local Ships' Environment*

Some of the noise contributions from the local ships' environment are as follows:

1 Atmospheric Noise from Absorption – Absorbing atmospheric media, such as water vapor, precipitation particles and oxygen emit thermal noise that can be described in terms of antenna noise temperature. These effects were discussed at the beginning of this chapter.
2 Industrial Noise – Heavy electrical equipment tends to generate broadband noise that can interfere with sensitive receivers. Therefore, a high percentage of this noise originates as broadband impulsive noise from ignition circuits. Namely, the noise varies in magnitude by as much as 20 dB, depending on whether it is measured on a normal working day or on weekends and holidays when it is lower in magnitude.
3 Out of Band Emission from Radar – Ship borne and surveillance radars operating in pulse mode can generate out of band emission that can interfere with SES receivers. In general, such emissions can be suppressed by inserting waveguide or coaxial filters at the radar transmitter output.

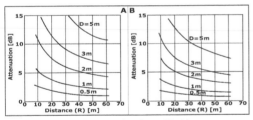

Figure 11. Estimated Attenuation due to Blocking

Courtesy of Handbook: "Radiowave Propagation Information for Predictions for Earth-to-Space Path Communications" by ITU

4 Interference from High Power Communication Transmitters – High power ships and terrestrial transmitters, for example HF ship radio transceivers; HF radio diffusion and TV broadcasting can interfere with SES.
5 Interference from Vehicles – Under certain operational conditions, RF emissions from vehicles may impair Rx sensitivity. Accordingly, in one measurement the noise emanating from heavy traffic has to be about –150dB (mW/Hz) within the frequency band 1.535 to 1.660 MHz.

6 Shipyard Noise – Extremely high peak amplitudes of noise of –141 dB (mW/Hz) were recorded from Boston Navy Yard, which was in full operation at that time. Thus, this noise is also a combination of city ambient noise and broadband electromagnetic noise from industrial equipment.

9.2 *Blockages Caused by Ship Superstructures*

Ship's superstructures can produce both reflection multipath and blockage in the direction of the satellite. For the most part, reflections from the ship's superstructure located on the deck can be considered coherent with the direct signal. The fading depth due to these reflections depends on a number of construction parameters including shape of the ship, location of the ship's antenna, antenna directivity and sidelobe level, axial ratio and orientation of the polarization ellipse, azimuth and elevation angles towards the satellite, etc. Antenna gain has a significant influence on the fading depth. In this case, low gain antennas with broader beam widths will collect more of the reflected radio signals, producing deeper fades.

Blockage is caused by ship superstructures, such as the mast and various types of antennas deployed on the ship. The geometry of blockage by a mast is presented in Figure 10. Signal attenuation depends on several parameters including diameter of column, size of antenna and distance between antenna and column. Accordingly, estimated attenuation due to blocking by a column type structure is shown in Figure 11 for antenna gains of 20 dB (A) and 14 dB (B), respectively.

9.3 *Motion of Ship's Antenna*

The motion of mobile satellite antennas is an important consideration in the design of MMSC systems. The received signal level is affected by the antenna off-beam gain because the antenna motion is influenced by the ship's motion. The random ship motion must be compensated by a suitable stabilizing mechanism to keep the antenna properly pointed towards the satellite. This is normally achieved either through a passive gravity stabilized platform or an active antenna tracking system. In either case, the residual antenna pointing error can be significant enough to warrant its inclusion in the overall link calculation.

Earlier experimental evidence suggests that the roll motion of a ship follows a zero mean Gaussian distribution over the short-term of the sea waves. The standard deviation of the distribution (σ_s) is a function of the vessel characteristics and the sea state of the wave height. In Figure 12 is illustrated the distribution of the instantaneous roll angle of a ship under moderate to rough sea conditions. The distribution of the ship motion approximates to a

Gaussian standard deviation of distribution with $\sigma_s = 5.42$ value.

Also shown in the figure is the distribution of roll angle of a passively stabilized antenna under the same conditions, which also follows a zero mean Gaussian distribution with a quantum of $\sigma_s = 0.99$. Solid curves in Figure 5.17. represent measured values and dashed curves show calculated values for stabilized antenna motion over the sea conditions with wave heights of approximately 5 m.

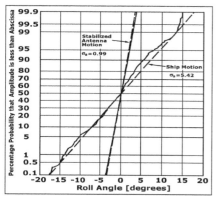

Figure 12. Measured Stabilized Antenna Motion
Courtesy of Handbook: "Radiowave Propagation Information for Predictions for Earth-to-Space Path Communications" by ITU

Otherwise, the relation between the standard deviations of the two distributions depends on the design of the passive stabilizer. Although the ship's antenna motion is much reduced, depending on the antenna beam width, the residual pointing error may be large enough to produce appreciable signal fluctuations. Over a long period of sea waves time σs varies as a function of the sea surface conditions and its distribution can be approximately by either a log normal distribution or a Weibull distribution.

10 CONCLUSION

The common satellite channel environment affects radiowave propagation in changeless ways. The different parameters influenced are mainly path attenuation, polarization and noise. The factors to be considered are gaseous absorption in the atmosphere, absorption and scattering by clouds, fog, all precipitation, atmospheric turbulence and ionospheric ef-

fects. In this sense, several measurement techniques serve to quantify these effects in order to improve reliability in the system design. Because these factors are random events, GMSC system designers usually use a statistical process in modeling their effects on radiowave propagation. To design an effective GMSC model it is necessary to consider the quantum of all propagation characteristics, such as signal lost in normal environment, path depolarization causes, transionospheric contribution, propagation effects important for mobile systems, including reflection from the Earth's surface, fading due to sea and land reflection, signal blockage and to the different local environmental interferences for all mobile and handheld applications. At any rate, the local propagation characteristics on the determinate geographical position have very specific statistical proprieties and results for ships, vehicles and aircraft.

REFERENCES

[01] Evans B.G., "Satellite communication systems", IEE, London, 1991.
[02] Freeman R.L., "Radio systems design for telecommunications (1-100 GHz)", John Wiley, Chichester, 1987.
[03] Fujimoto K. & other, "Mobile antenna systems handbook", Artech House, London, 1994.
[04] Galic R., "Telekomunikacije satelitima", Skolska knjiga, Zagreb, 1983.
[05] Group of authors, "Handbook - Mobile Satellite Service (MSS)", ITU, Geneva, 2002.
[06] Group of authors, "Handbook on Satellite Communications", ITU, Geneva, 2002.
[07] Group of authors, "Morskaya radiosyaz", Transport, Leningrad, 1985.
[08] Group of authors, "Radiowave propagation information for predictions for earth-to-space path communications", ITU, Geneva, 1996.
[09 Ilcev D. St. "Global Mobile Satellite Communications for Maritime, Land and Aeronautical Applications", Book, Springer, Boston, 2005.
[10] Maral G. & other, "Satellite communications systems", John Wiley, Chichester, 1994.
[11] Novik L.I. & other, "Sputnikovaya svyaz na more", Sudostroenie, Leningrad, 1987.
[12] Ohmori S. & other, "Mobile satellite communications", Artech House, Boston–London, 1998.
[13] Richharia M., "Mobile Satellite Communications – Principles and Trends", Addison-Wesley, Harlow, 2001.
[14] Sheriff R.E. & other, "Mobile satellite communication networks", Wiley, Chichester, 2001.
[15] Zhilin V.A., "Mezhdunarodnaya sputnikova sistema morskoy svyazi – Inmarsat", Sudostroenie, Leningrad, 1988.

18. Shipborne Satellite Antenna Mount and Tracking Systems

S. D. Ilcev
Durban University of Technology (DUT), Durban, South Africa

ABSTRACT: In this papers are introduced the very sensitive components of the ship's antenna tracking system as the weakest chain of the Maritime Mobile Satellite Service (MMSS). Also are presented the complete components of Ship Earth Station (SES), such as antenna system and transceiver with peripheral and control subsystems independent of ship motion. The communications Mobile Satellite Antennas (MSA) for Maritime Satellite Communications (MSC) are relatively large and heavy, especially shipborne directional Inmarsat B and Fleet-77 antenna systems. Over the past two decades the directional antenna system, which comprises the mechanical assembly, the control electronic and gyroscope, the microwave electronic package and the antennas assembly (dish, arrays and steering elements), is reduced considerably in both physical size and weight. These reductions, brought about be greater EIRP from satellite transponders coupled with GaAs-FET technology at the front end the receiver leading to higher G/N RF amplifiers, has made the redesigning, adopting and installing of shipborne antennas even on tracks and airplanes a reality.

1 INMARSAT MARITIME SHIP EARTH STATION (SES)

Inmarsat started operations 1981 with only Inmarsat-A maritime service and for that reason devised an initial synonym: INternational MARitime SATellite (INMARSAT). The SES is electronic equipment consisting in antenna and transceiver with peripheral devices usually installed on board ships or seaplatforms and rigs.

Later, Inmarsat developed other SES standards with mandatory and obligatory equipment, such as B, M, mini-M, C, mini-C, D, D+, Fleet F77, F55 and F33, including latest FleetBroadband. In Figure 1. are presented Above Deck Equipment (ADE) and Below Deck Equipment (BDE) general block diagrams of electronic units for SES terminals. The main elements of BDE are the following units [01]:

1 Cabin Interface Unit (CIU) – With build in PC or system processor Cabin Interface Unit controls and monitors the whole system operations of the transceiver and direction of the antenna dish, and also performs different task of baseband signal conditions between all obligation and optional BDE peripheral equipment (Tel, Fax, Tlx, Video on one hand and Data) and Baseband Processor on the another hand.
2 Baseband Processor – This processor simply performs baseband processing of all transmitting and receiving audio, video and PC or data signals. In such a way, Baseband Processor comprises Intermediate Frequency (IF) amplifier, modems and timing circuits for multiplexing up and down signals. The synthesizer produces the highly stable frequencies required for modulation and demodulation and for signal switching.
3 Interface Terminal – Connects navigation Gyro Compass and Motion Sensor Units with CIU device for satellite tracking and electronic control of ADE.

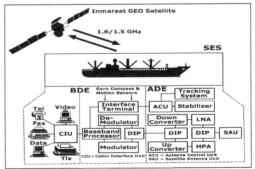

Figure 1. Maritime ADE and BDE Configuration
Courtesy of Book: "Global Mobile Satellite Communications" by St. D. Ilcev [01]

4 Modulator and Demodulator – Both represents the nucleus of any transceiver. First modulates the

baseband signal onto the IF carrier of transmitting signal and second demodulates the baseband signal from the IF carrier of the receiving signal.

5 BDE Diplexer (DIP) – Enables direction of transmitting signals from Modulator to the ADE and receiving signals from ADE to the Demodulator.

The ADE is mounted below the waterproof container or radome on the Stabilized Platform. The following units compose the ADE assembly:

1 ADE Diplexers (DIP) – The first diplexer passes all transmitting signals to the Up Converter and from Down Converter to the Demodulator. The second diplexer guides transmitting signals from HPA to the SAU and from SAU to the LNA.

2 Up and Down Converters – This unit accepts the modulated IF carrier from modulator and translates it to the uplink transmitted RF via HPA, by mixing with Local Oscillator (LO) frequency, while the Down Converter receives the modulated RF carrier from the LNA and translates its downlink receiving RF to the IF.

3 High Power Amplifier (HPA) – This unit provides amplification of transmitting signals by the Traveling Wave Tube Amplifier (TWTA) and Klystron Amplifier. The second HPA enables higher gain and better efficiency than TWTA but in smaller bandwidth of 2%. The amplified uplink signal goes via DIP to the SAU.

4 Low Nose Amplifier (LNA) – The LNA device provides initial amplifier stage of downlink signal coming from SAU via DIP without introducing much additional temperature noise. In this sense, the two most commonly used LNA products are new GaAs (Gallium Arsenide) FET (Field Effect Transistor) and old Parametric amplifiers. Thus, the recent developed GaAs FET LNA enables very low noise temperatures and takes advantages of its stability, reliability and low cost.

5 Antenna Control Unit (ACU) – Antenna Control Unit is providing control of the ship antenna Stabilized Platform (Stabilizer) and Tracking System. Thus, it maintains the antenna direction towards the focus of satellite against any motion of the ship.

2 ANTENNA MOUNT SYSTEMS

The MSA system is generally mounted on a platform, which has two horizontally, stabilized axes (X and Y), achieved by using a gyrostabilizer or sensors such as accelerometer or gyrocompass units. The stabilized platform provides a horizontal plane independent of mobile motion such as roll or pitch. For example, all mobiles have some kind of motions, but ship motion has seven components during navigation such as: roll, pitch, yaw, surge, sway, heavy, and turn, shown in Figure 2. Turn means a change in ship heading, which is intentional motion, not

caused by wave direction, and the other six components are caused by wave motion. Surge, sway, and heave are caused by acceleration.

2.1 Two-Axis Mount System (E/A and Y/X)

An antenna mount is mechanically moving system that can maintain the antenna beam in a fixed direction. In MMSS, the mount must have a function to point in any direction on the celestial hemisphere, because ships have to sail across the heavy seas. It is well known that the mount of the two-axis antenna configuration is the simplest mount providing such functions [02]

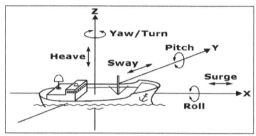

Figure 2. Components of Ship Motion
Courtesy of Book: "Mobile Antenna Systems Handbook" by K. Fujimoto and J.R. James [02]

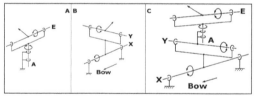

Figure 3. Two and Four-axis Mount Systems
Courtesy of Book: "Mobile Antenna Systems Handbook" by K. Fujimoto and J.R. James [02]

There are 2 typical mounts of the-axis configuration: one is E/A (elevation/azimuth) mount and the other is the Y/X mount. Simplified stick diagrams of both mounts are given Figure 3. (A) and (B) respectively. Thus, in the E/A mount, a full steerable function can be obtained by choosing the rotation range of the azimuth axis (A-axis) from 0-90°. In the Y/X mount a full steerable function is achieved by permitting the rotation angle from –90° to +90° to both the X and Y-axis. In fact, this is the basic configuration for the ships utility, so a special function required for its antenna mount system is to compensate the ship motions due to sailing and ocean waves, and to keep the antenna beam in nearly a fixed direction in space. In the case of the pointing and tracking under ship motions, the required rotation angle range of each axis is from 0° to more than 360° for the A-axis, and from –25° to +120° for the E-axis with respect to the deck level, assuming that the operational elevation angle is de facto restricted above 5° [01]. Otherwise, both mount types have several disadvantages.

2.2 Three-Axis Mount System (E/A/X, E'/E/A and X'/Y/X)

The three-axis mounting system is considered to be modified two-axis mount, which has one additional axis. The three-axis mount of an E/A/X type is shown in Figure 4. (A), which is the E/A mount with one additional X-axis. The function of the X-axis is to eliminate the rapid motion of the two-axis mount due to roll. However, in this system, the possibility of the gimbal lock for pitch is still left near the zenith, when the E-axis is parallel to the X-axis. The three-axis mount of an E'/E/A type shown in Figure 4. (B) is the E/A mount with an additional cross-elevation axis E. In the mount system, the charge of the azimuth angle is tracked by rotating the A-axis, and the change of the azimuth angle is tracked by a combined action of the E and E' axes. Hence, the E and E' axes allow movements in two directions at a right angle. With an approximate axial control, this mount is free from the gimbal lock problem near both the zenith and the horizon. The three-axis mount of an X'/Y/X type is the two-axis Y/X mount system with the X'-axis on it to remove the gimbal lock at the horizon, presented in Figure 4. (C). When the satellite is near the horizon, the X-axis takes out the rapid motion due to yaw and turn. In this sense, the X'-axis rotate within ± 120o, so the X'-axis can only eliminate the rapid motion within the angular range. In general, this axis mount is rather more complex than that of four-axis mount, because steering and stabilization interact with each other [02].

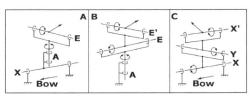

Figure 4. Three-axis Mount System
Courtesy of Book: "Mobile Antenna Systems Handbook" by K. Fujimoto and J.R. James [02]

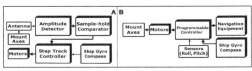

Figure 5. Functional Block Diagrams of Step and Program Tracking
Courtesy of Book: "Mobile Antenna Systems Handbook" by K. Fujimoto and J.R. James [02]

2.3 Four-Axis Mount (E/A/Y/X)

The stabilized platform made by X/Y-axis to take out roll and pitch, and two-axis mount of the E/A type is settled on the stabilized platform. This is four-axis mount solution, illustrated in Figure 3. (C). The tracking accuracy of this mount is the best solution because the stabilization function is separated from the steering function, and at any rate, four major components such as roll, pitch, and azimuth and elevation angles are controlled by its own axis individually as well. Accordingly, the four-axis mount has been adopted in many SES antenna systems of the current Inmarsat-A and B standards [01].

3 ANTENNA TRACKING AND POINTING SYSTEMS

Tracking and pointing system is another important function required of the antenna mount system. It should be noted that the primary requirement for SES tracking MSA systems are that they have to be economical, simple and enough reliable. Tracking performance is a secondary requirement when an antenna beam width is broad [02].

1 Manual Tracking - This is the simplest method, wherein an operator controls the antenna beam to maximize the received signal level. At first, the operator acquires the signal and moves the antenna around one axis of the mount. If the signal level increases, the operator continues to move the antenna in the same direction. If the signal decreases, the operator reverses the direction and continues to move the antenna until the signal level is maximized. The same process is repeated around the second axis and the antenna is held after both axes when the received signal level decreases. This method is suitable for Land Mobile Satellite Communications (LMSC) and especially for portable and fly-away terminals.

2 Step Tracking - Among various existing auto track systems, the step track system has recently been recognized as a suitable antenna-tracking mode for SES terminals because of its simplicity for moderate tracking accuracy. In such a way, recent design and development of integrated circuits and microprocessors have brought a greatly remarkable cost reduction to the step track system, which principle is the same as that of the manual track. The only difference is that an electric controller plays the role of an operator in the manual track. The schematic block diagram of the step tracking system is shown in Figure 5. (A). Sample-hold circuits are used to hold the signal level, which are compared before and after the antenna has been moved by a present angular step. If the level is increasing, the antenna is moved in the same direction, and vice versa, if

the level is decreasing, the direction will be reversed. This process will be carried out alternately between two axes levels, which accuracy depends on the sensitivity of comparators. As a result, the beam center is maintained in the vicinity of the satellite direction. Thus, wrong decisions on the comparison of levels generally arise from the S/N ratio and the level changes due to the multipath fading and the stabilization error.

3 Program Tracking - The concept of the program system is based on the open-loop control slaving to the automatic navigation equipment, such as a ship gyrocompass, GPS, the Omega and Loran-C systems. In the program tracking, the antenna is steered to the point of the calculated direction based on the position data of the navigation equipment. Since the satellite changes because of roll, pitch, and turn direction, a function to remove the rapid motions is required in the program track, which block diagram is shown in Figure 5. (B). The error of navigation equipment is negligibly small for the program track system, while the error of it mainly depends on the accuracy of sensors for roll, pitch, and turn directions, what is the stabilization error. In fact, an adequate sensor for the program track system is a vertical gyro, because it is hardly affected by the lateral acceleration. When the stabilization requirement is lenient, the conventional level sensor, such as inclinometer, a pendulum, and a level, may be used with careful choice of the sensor's location.

Therefore, the controller calculates the direction of the satellite orbit to compensate differences for the ship's motions affected by all components. In any event, the simpler the axis configuration of the mounts, the more complex the program calculation procedure becomes.

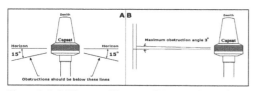

Figure 6. Safe Distance of Inmarsat-C Antenna from Obstructions
Courtesy of Manual: "Sailor Maritime Inmarsat-C" by Thrane & Thrane [03]

More exactly, since the program controller has to execute different calculations of many trigonometric functions, a microprocessor is a candidate for the controller. However, the program tracking system is also applicable to the four-axis mount. A combination with the step track system is more desirable because the error of the program track system can be compensated by the step track system and its error

due to the rapid ship motions can be compensated for by the program tracking system [01].

4 OMNIDIRECTIONAL SHIPBORNE MSA MOUNTING

When installing MSA is necessary to find a location on board of ship that is as free from any obstructions as possible. On the other hand, also is important to maintain a certain distance to other communication antenna systems, especially radar installations. Finally, the best place for the MSA on board ship would be above radar scanning antennas or far a way from them. Otherwise, the minimum safe distance should be maintained to HF antenna 5 m, to VHF antenna 4 m, and to magnetic compass 3 m [04].

The omnidirectional antenna is designed to provide satellite coverage even when the vessel has pitch and roll movement up to 15°. In this sense, to maintain this coverage the ship antenna should be free from any obstructions in the area down to 15° below the horizon, as is shown in Figure 6. (A). Since this may not be possible in the fore and aft directions of the vessel, the clear area can be reduced to 5° below the horizon in the fore and aft directions and 15° below the horizon in the port and starboard directions. Otherwise, any compromise in this recommendation will degrade performance. If an obstruction such as a pole or a funnel is unavoidable, the distance to these objects should large enough, so that the obstruction only covers 3o. For instance, if the diameter of ship obstruction object is 0,1 m, the safe distance should be about 2 m, as is shown in Figure 6. (B).

The safety levels for the Thrane & Thrane Capsat-C Antenna Unit and similar Inmarsat-C aerials are based on the ANSI standard C95.1-1982. Namely, this standard recommends the maximum power density at 1,6 GHz exposed to human beings not to exceed 5 mW/cm^2. Therefore, at the maximum radiation output power from Inmarsat-C antenna of 16 dBW EIRP corresponds to a minimum safety distance of about 30 cm. So, the future standard from the European Telecommunication Standard Institute (ETSI) concerning 1,5/1,6 GHz Mobile Earth Station (MES) the new recommendation will be maximum 8 W/m^2 (0,8 mW/cm^2) with minimum safety distance on 62 cm at 16 dBW of EIRP [01].

5 DIRECTIONAL MSA MOUNTING AND STEERING

The directional ship Above Deck Equipment (ADE) consists of an antenna unit mounted on a pedestal, an RF unit, power and control unit, all covered by a radome. Ideally, the antenna should have free optical sight in all directions above an elevation angle of 5o.

The antenna must be placed as high as possible on the best position on board of ship to avoid blind spots with degradation and/or loss of communication link, caused by different deck obstacles [05].

5.1 Placing and Position of SES Antenna Unit

The directional antenna has a beamwidth of 10° and ideally requires a free line of sight in all directions above an elevation angle of 5°. Possible obstructions will cause blind spots, with the result of degradation or even loss of communication link with the satellite. So, complete freedom from degradation of the signal propagation is only accomplished by placing the ship antenna above the level of possible obstructions.

Figure 7. Theoretical Antenna Installation
Courtesy of Book: "Global Mobile Satellite Communications" by St. D. Ilcev [01]

This is normally not feasible and a compromise must be made to reduce the amount of blind spots. The degree of degradation of the communication depends on the size of the ship obstructions as seen from the antenna, hence the distances to them must also be considered. However, it should be remembered that the antenna RF beam of energy possesses a width of 12° angle cone and consequently, objects within 10 m of the radome, which cause a shadowing sector greater than 6°, are not likely to degrade the electronic equipment significantly. Preferably, all obstructions within 3 m of the shipborne antenna system should be avoided.

Obstructions less than 15 cm in diameter can be ignored beyond this distance. Knowing the route that the ship normally sails allows a preferable sector of free sight to be established, thus facilitating the location of the antenna unit.

In such a way, the antenna beam must be capable of being steered in the direction of any GEO satellite of the Inmarsat constellation, whose orbital inclination does not exceed 3° and whose longitudinal excursions do not exceed ±0.5°. Therefore, means must be provided to point the antenna beam automatically towards the satellite with sufficient accuracy to ensure that the G/T and EIRP requirements, namely receive and transmit signal levels, are satisfied continuously under operational conditions.

Careful and important consideration should be given to the placing of an Inmarsat-A or B standard ADE with antenna radome. Essentially, the focal point of the parabolic antenna must be pointing directly at the GEO satellite being tracked without any interruption of the microwave beam, which may be caused by any obstruction on the ship. Inmarsat specify that there should be no obstacle that is likely to downgrade the performance of the equipment in any angle of azimuth down to an elevation of –5°, which is not easy to achieve. Thus, the SES design and installation guidelines of Inmarsat explains a theoretical satellite antenna installation instructions mode satisfying this advice but with the disadvantage that the antenna system is very high above the vessel's deck and would be impossible to install in such a way, see Figure 7. Differently to say, this type of installation is not practical because in reality a ship's satellite antenna would certainly be very adversely affected by extremely strong and gusty wind, will have higher inclination angles, vibration and it would be difficult to gain access for maintenance purposes [06].

If ship's structures do interrupt the antenna beam, blind sectors will be caused, leading to degraded communications over some arc of azimuth travel. Thus, if it is like that and as is often the case, it is impossible to find a mounting position free from all obstructions; the identified blind sectors should be recorded. It may be possible for the operator, when in an area served by two satellites, to select the one whose azimuth and elevation angles with respect to the ship's position are outside the blind sector. Obviously, this method is not enough practical because the satellite overlapping sectors inside of Inmarsat's four ocean regions cover relatively small areas. The best solution to avoid all blind sectors is to place the antenna unit on top of the radar mast or on a specially designed mast.

5.2 Antenna Mast and Stabilizing Platform

The mast has to be designed to carry the weight of the antenna unit, maximum 300 kg, depending on model design or manufacturer, presented in Figure 8. (A). It must also be able to withstand the forces imposed by severe winds up to 120 knots on the radome and strong vibrations due to very rough seas on the whole ADE construction. The top end of the mast should be fitted with a flange with holes matching the bolts extending from the bottom of the radome. In addition, the flange must not be so large as to interfere with the hatch in the bottom of the antenna unit. In this sense, the holes through the mast flange must be positioned symmetrically around the ship's longitudinal axis.

If the height of the mast makes it necessary to climb up to the antenna unit, a ladder must be provided on the mast column. A guardrail must be attached to the upper section for safety purposes. Finally, if the height of the mast exceeds approximately 4.5 m, an access platform should be attached to the mast about 1.5 m below the radome bottom.

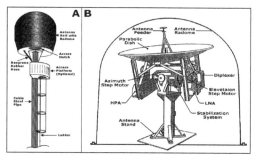

Figure 8. ADE Mast and Stabilized Platform
Courtesy of Manuals: "Saturn 3" by EB Communications and "Maritime Communications" by Inmarsat [04]

The radome completely encloses the periphery of the base plate assembly to protect the electronic and mechanical components from corrosion and weather. It is usually fabricated from high-gloss fibreglass and is electronically transparent to RF signals in the assigned frequency band. The radome is secured to the circular or square antenna stand (base plate) with several screws and can be removed easily without special tools.

The antenna-stabilized platform is housed inside the radome and consists in the electrical and mechanical elements, presented in Figure 8. (B). At this point, there are two antenna control stepper motors. First, there is the azimuth step motor, which controls the position of the antenna reflector in the horizontal A plane (azimuth) and second, is the elevation step motor that controls the vertical E plane (elevation). Each motor has four phase-inputs coming from the drive circuit in the BDE Control Board and a supply voltage from the master power supply located on the antenna stand. Accordingly, as a stepper motor turns the antenna, it also adjusts the setting of the relevant sensor potentiometer. The sensor voltage supply is the reference voltage for the A/D converter on the Control Board. In the other words, two stepper motors move the satellite antenna in both azimuth and elevation angles and moves the relevant sensor potentiometers, which provide feedback information on the position of the antenna. In fact, the antenna stabilization system, or gyroscope with two gyro motors, stabilizes the platform for the antenna against the roll and pitch of the ship. A two-turn solenoid clamps the antenna platform to the gyroscope assembly. Therefore, a diplexer passes the Rx signal from the antenna to the LNA and the Tx signal from the transceiver assembly to the antenna. The LNA amplifies the Rx signal and the HPA amplifies the Tx signal. The parabolic dish radiates EM energy to and from the antenna feeder in Rx or Tx direction, respectively. In such a mode, the antenna assembly

is mounted below the radome for protection purposes [07].

5.3 Antenna Location Aboard Ship

The ship's antenna unit should be located at a distance of at least 4.5 m from the magnetic steering compass. At this point, it is not recommended to locate the antenna close to any interference sources or in such a position that sources such as the radar antenna, lie within the antenna's beam width of 10o when it points at the satellite. The ADE should also be separated as far as possible from the HF antenna and preferably by at least 5 m from the antenna systems of other communications or navigation equipment, such as the antenna of the satellite navigator or the VHF and NAVTEX antennas. In addition, it is not practical to place the antenna behind the funnel, as smoke deposits will eventually degrade antenna performance. Regardless of the location chosen for the antenna, it should be oriented to point forwards in parallel with the ship's longitudinal axis when in the middle of its azimuth range, which will correspond to zero degrees on the azimuth indicator [08].

The EM RF signals are known to be hazardous to health at high radiation levels. In such a way, it is inadvisable to permit human beings to stand very close to the radome of an SES when the system is communicating with a satellite at a low elevation angle. In this case, Inmarsat recommends that the radiation levels in the vicinity of the antenna should be measured. The crew members and passengers should not be admitted to areas closer than 10 m away from the antenna unit at desk level above 2 m, measured beneath the lowest point of the radome, as shown in Figure 9. (A).

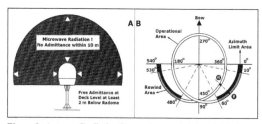

Figure 9. Antenna Radiation Precautions and Azimuth Limit
Courtesy of Manual: "Saturn 3" by EB Communications [04]

No restrictions, therefore, are required when the antenna radome is installed at least 2 m above the highest point accessible to crew and passengers. Authorized personnel should not remain close to the antenna system for periods exceeding 1 hour per day without switching off the RF transmitter. However, radiation plan diagrams may be produced and located near the antenna as a warning for crew members, passengers and ship's visitors, or distances from the antenna may be physically labeled at the relevant place.

5.4 Satellite Determination and Antenna Azimuth Limit

An Inmarsat-A, B and M MSA must be capable of locating and continuously tracking the GEO satellite available or selected for communication, namely if the ship has in view only one satellite or if the ship is in an overlapping position, respectively. Thus, Inmarsat-C has an omnidirectional antenna and does not need a tracking system. Locating and tracking may be done automatically, as in the case of an SES, or manually, as with a portable MES [01].

In fact, it is common practice to believe that the GEO satellites are fixed and that once the link has been established it will remain so as long as the mobile does not move. However, ships or other mobiles are always moving during operational management of voyages and satellites are under the influence of a number of variable astrophysical parameters, which cause it to move around its station by up to several degrees. At this point, an ADE tracking system must counteract this by repositioning the SES antenna at regular intervals and in case of need. The carrier signal is monitored continuously and, if a reduction in its amplitude is detected, a close-programmed search is initiated until the carrier strength is again at maximum. No loss of signal occurs during this process, which is automatically initiated. Obviously, the greatest tracking problem will arise when the SES is moving at speed with respect to the satellite. In a more general sense, an Inmarsat-A and B MSA may be moved through any angle in azimuth and elevation as the vessel moves along its course. In this case, it is essential that electronic control of the antenna is provided. In practice, ship antenna control system may be achieved by manual and/or even automatically by simple electronic feedback methods [02].

1 Manual Commands – When the radio or navigation operator onboard ships has selected manual control, elevation is commanded by up and down keys, whereas azimuth positioning is controlled clockwise and counter-clockwise keys. In such a case, a command would be used when the relative positions of both the vessel and the satellite are known. Azimuth and elevation angles of antenna can be derived and input to the equipment, by using the two A and E charts of Inmarsat satellite network coverage. Once the antenna starts to detect a satellite signal, the operator display indicates signal strength. Fine positioning can now be achieved by moving the ship antenna in A and E in 1/6th degree increments until maximum field strength is achieved.

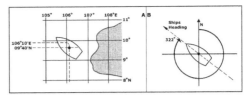

Figure 10. Antenna Pointing
Courtesy of Manual: "Saturn 3" by EB Communications [04]

2 Automatic Control – Once geostationary satellite lock has been achieved, the system will automatically monitor signal strength and apply A/E corrections as required in order maintaining this lock as the vessel changes course.

3 Automatic Search – An automatic antenna search routine commences 1.5 minutes after switching on the equipment, or it may be initiated by the operator. Therefore, the elevation motor is caused to search between 5° and 85° limits, whereas the azimuth motor is stepped through 10° segments. In s such a manner, if the assigned common signaling channel signal is identified during this search the step tracking system takes over to switch the antenna above/below and to each side of the detecting signal location searching for maximum signal strength.

4 Gyroscopic Control – Using this mode the lock is maintained irrespective of changes to the vessel's course by sensing signal changes in the ship's gyro repeater circuitry. In the proper manner, satellite signal strength is monitored and if necessary, the A/E stepper motors are commanded to start searching for the maximum signal strength.

5 Antenna Rewind – The antenna in the ADE is provided on a central mast and is coupled by various control and signal cables to a stationary stable platform. Thus, if the antenna was permitted to rotate continuously in the same direction, the feeder cables would eventually become so tightly wrapped around the central support that they would either prevent the antenna from moving or they would fracture. To prevent this happening, a sequence known as antenna rewind is necessary, as is shown in Figure 9. (B). In fact, an antenna has three areas with rewind time of approximately 30 seconds plus stabilizing time, giving a total of about 1.5 minutes:

– Operational Area is the antenna-rotating limit in the azimuth plane. In fact, the antenna can rotate a total of 540°, which is shown as a white area in Figure 9. (B). Normally, the vessel antenna will operate in the operational area, which is between 60° and 480°.

– Rewind Area is necessary for the following reasons: if the antenna moves into one of the rewind areas, i.e., 10° to 60° or 480° to 530° (antenna azimuth lamp lights) and if no traffic is in progress, the antenna will automatically rewind 360° to get into the operational area and still be pointed at the satellite, which is illustrated as a dotted area in Figure 9. (B). For example, the antenna moves from position 1 to 2 and the rewind lamplights. If the SES is occupied with a call and the ship turns so that the antenna enters the rewind area, no rewind will take place unless the antenna comes into the azimuth limit area.

If this happens, rewind will take place and the call will be lost. The azimuth-warning indicator on the operator display will light to indicate that antenna rewinding is in progress.

– Azimuth Limit Area is an important factor because when the antenna is in this area the azimuth limit lamp lights. If the antenna moves into the outer part of the azimuth limit area, i.e., 0° to 10° or 530° to 540°, rewind will start automatically, despite traffic in progress.

5.5 *Antenna Pointing and Tracking*

The directional reflector antenna is highly directive and must be pointed accurately at the satellite to achieve optimum receiving and transmitting conditions. In normal operation the antenna is kept pointed at the satellite by the auto tracking system of, for example, Saturn 3 SES. Before the auto tracking can take over, the antenna must be brought within a certain angle in relation to the satellite. This can be obtained using the command "find" or by manually setting the antenna using the front push buttons on the terminal or via teleprinter command. For manual pointing it is necessary to provide the ship's plotted position, ship's heading by gyro, azimuth angle and elevation angle map of the satellite [01].

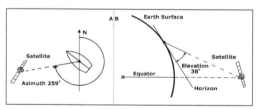

Figure 11. Azimuth and Elevation Angle
Courtesy of Manual: "Saturn 3" by EB Communications [04]

1   Ship's Plotted Position – The plotted position is needed to decide which satellite can be used, namely which Inmarsat network area can be tuned: Indian Ocean Region (IOR), Pacific Ocean Region (POR), Atlantic Ocean Region-West (AORW) or Atlantic Ocean Region-East (AORE), depending on the ship's actual position, as presented in Figure 10. (A). Sometimes, the ship can be in an overlapping area covered by two or even three Inmarsat satellites. In this case it will be important to choose convenient CES and

to point the antenna towards one of overlapping ocean regions.

2   Ship's Heading by Gyrocompass – The method of the permanent heading of the ship course determined by gyrocompass is needed for the antenna auto-tracking and focusing system during navigation, illustrated in Figure 10. (B). In this case, this scenario is not needed for omnidirectional ship antennas, such as for Standard-C, mini-C and D+ equipment. This connection is very important for new Inmarsat Fleet and FleetBroadband standards.

3   Azimuth Angle – This is the angle between North line and horizontal satellite direction as seen from the ship, as is shown by example of 259o, in Figure 11. (A). The actual azimuth angle for the various satellites due to the ship's plotted position can be found on the azimuth angle map.

4   Elevation Angle – The elevation angle is the satellite height above the horizon as seen from the ship, as is shown by the example of 38o, in Figure 11. (B). In this case, the actual elevation angle for the various satellites due to the ship's plotted position can be found on the Elevation angle map.

6 CONCLUSION

It is obvious that shipborne MSA configuration needs to be compact and lightweight. These requirements will be difficult to achieve because ship's directional antenna has quite heavy components for stabilization and tracking, and because the compact antenna has two major electrical disadvantages such as low gain and wide beam coverage. Therefore, a new generation of powerful satellite transponders with high EIRP and G/T performances should permit the effective design of more powerful, compact and lightweight MSA for ship applications. On the other hand, new physical shapes of radome and less weight of components are very important requirements in connection with compactness and lightweight, what will permit easier installation and regular maintenance of ship antennas. With shipborne antennas on oceangoing large ships, installation requirements are not as limited compared to fishing vessels or very small boats and yachts, because even small ships have enough space to put mast on compass deck and to install an antenna system. However, in the case of small ships, especially yachts, very low profile and lightweight equipment is required, such as the Inmarsat Fleet F33, Mini-M or C omnidirectional antenna installations.

REFERENCES

[01] Ilcev D. St. "Global Mobile Satellite Communications for Maritime, Land and Aeronautical Applications", Book, Springer, Boston, 2005.

[02] Fujimoto K. & James J.R. "Mobile Antenna Systems Handbook", Artech House, Boston, 1994.

[03] Group of Authors, "Sailor Maritime Inmarsat-C Installation Manual", Thrane & Thrane, Soeborg, 1997.

[04] Group of Authors, "Saturn 3 standard-A Installation/Operator' Manuals", EB, Nesbru, 1986.

[05] Evans B.G., "Satellite Communication Systems", IEE, London, 1991.

[06] Gallagher B. "Never Beyond Reach", Book, Inmarsat, London, 1989.

[07] Rudge A.W., "The Handbook of Antenna Design", IEE, London, 1986.

[08] Law E.P. "Shipboard Antennas", Artech, 1983.

19. Yesterday, Today and Tomorrow of the GMDSS

K. Korcz
Gdynia Maritime University, Gdynia, Poland

ABSTRACT: Basing on a general concept, the main functions and the international requirements the Global Maritime Distress and Safety System (GMDSS) have been presented. The modifications of the system since its implementation and its current status have been described. The future of the GMDSS has been discussed as well.

1 INTRODUCTION

In 1988, the Conference of Contracting Governments to the 1974 SOLAS Convention on the Global Maritime Distress and Safety System (GMDSS) adopted amendments to the 1974 SOLAS Convention concerning radiocommunications for the GMDSS. These amendments entered into force on 1 February 1992. On 1 February 1999 the GMDSS has become implemented for all SOLAS ships.

Since full implementation of the GMDSS some changes both of regulatory and technical nature have occurred.

The Maritime Safety Committee (MSC) at its 81st session decided to include, in the work programmes of the NAV and Radiocommunications and Search and Rescue (COMSAR) Sub-Committees, a high priority item on "Development of an e-navigation strategy". One of the fundamental elements of e-navigation will be a data communication network based on the some GMDSS infrastructure elements. It follows that question on modernization of the GMDSS is legitimate.

2 THE ORIGINAL CONCEPT OF THE GMDSS

The original concept of the GMDSS is that search and rescue authorities ashore, as well as shipping in the immediate vicinity of the ship in distress, will be rapidly alerted to a distress incident so they can assist in a coordinated search and rescue operation with the minimum delay. The system also provides for urgency and safety communications and the promulgation of maritime safety information (MSI) (Czajkowski, 2000).

2.1 *Functional requirements*

The GMDSS lays down nine principal communications functions which all ships, while at sea, need to be able to perform (IMO, 2004):
1 transmitting ship-to-shore distress alerts by at least two separate and independent means, each using a different radiocommunication service;
2 receiving shore-to-ship distress alerts;
3 transmitting and receiving ship-to-ship distress alerts;
4 transmitting and receiving search and rescue co-ordinating communication;
5 transmitting and receiving on-scene communication;
6 transmitting and receiving signals for locating;
7 transmitting and receiving maritime safety information;
8 transmitting and receiving general radiocommunication from shorebased radio systems or networks;
9 transmitting and receiving bridge-to-bridge communication.

2.2 *Radiocommunication services*

The following radio services are provided for the GMDSS:
– a radiocommunication service utilizing geostationary satellites in the maritime mobile satellite service (INMARSAT);
– a radiocommunication service utilizing polar orbiting satellites in the mobile satellite service (COSPAS-SARSAT);
– a radiocommunication service for transmitting signals from survival craft stations in the 9200 – 9500 MHz band;

– the maritime mobile service in the bands between 156 MHz and 174 MHz (VHF);
– the maritime mobile service in the bands between 4,000 kHz and 27,500 kHz (HF);
– the maritime mobile service in the bands 415 kHz to 535 kHz and 1,605 kHz to 4,000 kHz (MF).

2.3 GMDSS Sea areas

Radiocommunication services incorporated in the GMDSS system have individual limitations with respect to the geographical coverage and services provided. The range of communication equipment carried on board the ship is determined not by the size of the ship but by the area in which it operates. Four sea areas for communications within the GMDSS have been specified by the IMO. These areas are designated as follows (IMO, 2004):
– Sea area A1 – an area within the radiotelephone coverage of at least one VHF coast station in which continuous DSC alerting is available.
– Sea area A2 – an area, excluding sea area A1, within the radiotelephone coverage of at least one MF coast station in which continuous DSC alerting is available.
– Sea area A3 – an area, excluding sea areas A1 and A2, within the coverage of an INMARSAT geostationary satellite in which continuous alerting is available.
– Sea area A4 – an area outside sea areas A1, A2 and A3 (the polar regions north and south of 75° latitude, outside the INMARSAT satellite coverage area).

2.4 Equipment carriage requirements

Based on the range limitations of each radiocommunication service the four sea areas have been defined according to the coverage of VHF, MF, HF Coast Radio Services and Inmarsat services. The type of radio equipment required to be carried by a ship is therefore determined by the sea areas through which a ship travels on its voyage.

2.5 GMDSS equipment and systems

The following equipment and systems are provided for the GMDSS (Fig. 1):
– DSC - Digital Selective Calling;
– INMARSAT Satellite System;
– EPIRB - Emergency Position Indicating Radio-Beacon (Inmarsat E, Cospas/Sarsat and VHF DSC);
– SARTs - Search And Rescue Transponders;
– NAVTEX System;
– NBDP - Narrow Band Direct Printing;
– RTF - Radiotelephony;
– DMC - Distress Message Control;
– navigational equipment (for support).

Other elements of GMDSS to be showed in Fig. 1 stand for as follows:
– CES - INMARSAT *Coast Earth Station*;
– SES - INMARSAT *Ship Earth Station*;
– LUT - COSPAS/SARSAT *Local User Terminal*;
– RCC - *Rescue Coordination Centre*.

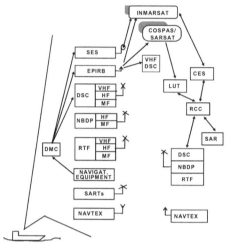

Figure 1. GMDSS equipment and systems (Korcz, 2007)

2.5.1 DSC specification

Digital selective calling (DSC) is designed for automatic station calling and distress alerting. Each call consists of a packet of a digitized information. DSC calls can be routed to all stations, to an individual station or to a group of stations.

The system is used by ships and coast stations in the MF, HF and VHF maritime communication bands.

The system is a synchronous system using characters composed from a ten bit error-detecting code. The first seven bits are information bits. The last three bits are used for error-detection. Each character is sent twice but separated in time and a message check character added at the end of the call.

Technical characteristics and operational procedures for the use of DSC equipment are described in the following documents:
– Recommendation ITU-R M.493 'Digital selective calling system for use in the Maritime Mobile Service'.
– Recommendation ITU-R M.541 'Operational procedures for the use of digital selective-calling (DSC) equipment in the Maritime Mobile Service'.

2.5.2 Inmarsat specification

The original concept of the GMDSS includes three Inmarsat services: A, B and C.

Inmarsat A provides two-way direct-dial phone connection as well as fax, telex and data services at rates between 9.6kbps up to 64kbps.

Inmarsat B was first maritime digital service, launched in 1993, and remains a core service for the maritime industry. It supports global voice, telex, fax and data at speeds from 9.6kbps to 64kbps, as well as GMDSS - compliant distress and safety functions.

Inmarsat C is one of the most flexible mobile satellite message communication systems, it has the ability to handle commercial, operational and personal messages just as easily as distress and safety communications.

It offers two-way, store-and-forward packet data communication via a lightweight, low-cost terminal. Inmarsat C is recommended for the any of the following applications:
- E-mail and messaging
- Fax and telex
- SMS text
- Remote monitoring
- Tracking
- Chart and weather updates
- Maritime safety information
- GMDSS
- SafetyNET and FleetNET

2.5.3 *NBDP specification*

The Narrow Band Direct Printing – **NBDP** (radiotelex) systems employs error correction in the form of ARQ (*Automatic Retransmissions Request*) and FEC (*Forward Error Correction*). The technical details of the error correction are defined by the ITU-R in Recommendation M.476 and the Recommendation M.625. Radiotelex is also known as Telex Over Radio (TOR).

2.5.4 *NAVTEX specification*

International **NAVTEX** (NAVigational TelEX) service means the co-ordinated broadcast and automatic reception on the frequency 518 kHz of maritime safety information (MSI) by means of Narrow Band Direct Printing (NBDP-FEC) telegraphy. The operational and technical characteristics of the NAVTEX system are contained in Recommendation ITU-R M.540. Performance standards for shipborne narrow-band direct-printing equipment are laid down in IMO Assembly resolution A.525(13).

The principal features of NAVTEX service are as follows:
- the service uses a single frequency (518 kHz) on which coast stations transmit information in English on a time-sharing basis to prevent mutual interference; all necessary information is contained in each transmission;
- the power of each coast station transmitter is regulated so as to avoid the possibility of interference between coast stations; Navtex transmissions provide a range of about 250 to 400 nautical miles;
- dedicated Navtex receivers are used on the board of the ships; they have the ability to select messages to be printed, according to a technical code (B1B2B3B4) which appears in the preamble of each message.

3 LAST DECADE GMDSS MODIFICATION

The last decade modification of the GMDSS has concerned both technical and regulatory issues.

3.1 *Technical modification of GMDSS*

The most important GMDSS modification has concerned the Inmarsat. In 1999, Inmarsat became the first intergovernmental organisation to transform into a private company and, in 2005, was floated on the London Stock Exchange. It caused that at present Inmarsat is recognised as a leader in mobile satellite communication field.

Inmarsat Fleet service provides both ocean-going and coastal vessels with comprehensive voice, fax and data communications. At present the Fleet range of services includes:
- Fleet 77
- Fleet 55
- Fleet 33

Inmarsat Fleet 77 has been introduced in 2002, and Inmarsat Fleet 55 and 33 in 2003.

Inmarsat Fleet's high-quality Mobile ISDN and cost-effective IP-based Mobile Packet Data Services offer unparallel connectivity for access to e-mail and the Internet, weather updates, video conferencing and vessel monitoring systems.

Fleet 33 offers global voice as well as fax and a choice of data communications at up to 9.6kbps.

Fleet 55 offers global voice and high-speed fax and data services at up to 64kbps.

Fleet 77 is Inmarsat's most advanced maritime service, providing global voice and high-speed fax and data services at up to 128kbps. It fully supports the GMDSS and includes advanced features such as emergency call prioritization, as stipulated by IMO Resolution A.1001 (25). Fleet F77 also helps meet the requirements of the International Ship and Port Facility Security (ISPS) code, which enables the cost-effective transfer of electronic notices of arrival, crew lists, certificates and records.

Inmarsat Fleet series are recommended for the applications showed in Table 1.

Table 1. Applications of Inmarsat Fleet series

	Fleet 33	Fleet 55	Fleet 77
Data transfer	+	+	+
Internet	+	+	+
E-mail and messaging	+	+	+
Fax	+	+	+
SMS text	+	+	+
Voice	+	+	+
Crew calling	+	+	+
Encryption	-	-	+
Videoconferencing	-	-	+
Remote monitoring	+	+	+
Weather updates	+	+	+
Telemedicine	+	+	+
GMDSS functions	-	-	+

Because *Fleet 77* is IP compatible, it supports an extensive range of commercially available off-the-shelf software, as well as specialized maritime and business applications. Fleet 77 also ensures cost-effective communications by offering the choice of Mobile ISDN or MPDS channels at speeds of up to 128kbps.

FleetBroadband is Inmarsat's next generation of maritime services delivered via the Inmarsat-4 satellites. It is commercially available since the second half of 2007. The service is designed to provide the way forward for cost-effective, high-speed data and voice communications (Table 2).

Users have the choice of two products (FB250 and FB500). Both use stabilized, directional antennas, which vary in size and weight. The above deck antennas are smaller than other Fleet products.

Table 2. FleetBroadband performance capabilities

	FB250	FB500
Data		
Standard IP	Up to 284kbps	Up to 432kbps
Streaming IP	32, 64, 128kbps	32, 64, 128, 256kbps
ISDN	–	64kbps
Voice	4kbps and digital 3.1kHz audio	
Fax	Group 3 fax via 3.1kHz audio	
SMS	Standard 3G (up to 160 characters)	
Antenna		
Diameter from	25cm	57cm
Height from	28cm	68cm
Weight from	2.5kg	18kg

FleetBroadband supports an extensive range of commercially available, off-the-shelf software, as well as specialized user applications. It is ideal for:
- Email and webmail
- Real-time electronic chart and weather updates
- Remote company intranet and internet access
- Secure communications
- Large file transfer
- Crew communications
- Vessel/engine telemetry
- SMS and instant messaging
- Videoconferencing
- Store and forward video.

It should be also noted that **Inmarsat E** service ceased to be supporting GMDSS in 2006 and **Inmarsat A** service – in 2007.

In the same time, instead of the Inmarsat E service, the new Cospas-Sarsat Geostationary Search and Rescue System (**GEOSAR**) has been introduced as completion of the Low-altitude Earth Orbit System (LEOSAR). These two Cospas-Sarsat systems (GEOSAR and LEOSAR) create the complementary system assists search and rescue operations (SAR).

As the result of the hard work of International Maritime Organization (IMO) other two new systems have been introduced:
- Ship Security Alert System – **SSAS** (in 2004),
- Long-Range Identification and Tracking of ships – **LRIT** (in 2009).

Although the above mentioned systems are not a part of the GMDSS, in the direct way they use its communication means.

At the end it's worth to note that the International Cospas-Sarsat System has ceased satellite processing of **121.5/243 MHz beacons** on 1 February 2009 and that on 1.01.2010 AIS-Search and Rescue Transmitter (**AIS-SART**) have been introduced as well.

3.2 Regulatory modification of GMDSS

At a regulatory level a modification of the GMDSS is coordinated by two international organizations: International Maritime Organization (IMO) and International Telecommunication Union (ITU).

IMO modifications are mainly concerning to the amendments to Chapter IV of the International Convention for the Safety of Life At Sea (SOLAS) and to the IMO resolutions.

From the Radiocommunication point of view, the most important modification was adoption by IMO of Resolution A.1001(25) dated 29.11.2007 on Criteria for the Provision of Mobile Satellite Communication Systems in the GMDSS and revision of Chapter IV of IMO SOLAS Convention extends the International Mobile Satellite Organization (IMSO) oversight to GMDSS Services provided by **any satellite operator** which fits criteria.

ITU modifications are mainly concerning to the amendments to Radio Regulations. These amendments were adopted by two Word Radiocommunication Conferences (**WRC**).

The first World Radiocommunication Conference took place in 2003 (WRC-03) and in the field of maritime radiocommunication it took up following main issues:
- to consider Appendix **13** and Resolution **331 (Rev.WRC-97)** with a view to their deletion and, if appropriate, to consider related changes to Chapter VII and other provisions of the Radio Regulations, as necessary, taking into account the continued transition to an introduction of the

Global Maritime Distress and Safety System (GMDSS) (Agenda Item 1.9);

– to consider the results of studies, and take necessary actions, relating to exhaustion of the maritime mobile service identity numbering resource (Resolution **344** **(WRC-97)**) (Agenda Item 1.10.1);

– to consider the results of studies, and take necessary actions, relating to shore-to-ship distress communication priorities (Resolution **348 (WRC-97)**) (Agenda Item 1.10.2);

– to consider measures to address harmful interference in the bands allocated to the maritime mobile and aeronautical mobile (R) services, taking into account Resolutions **207 (Rev.WRC-2000)** and **350 (WRC-2000)**, and to review the frequency and channel arrangements in the maritime MF and HF bands concerning the use of new digital technology, also taking into account Resolution **347 (WRC-97)** (Agenda Item 1.14).

The second Word Radiocommunication Conference took place in 2007 (WRC-07) and the main maritime radiocommunication items were as follows (ITU, 2008):

– taking into account Resolutions **729 (WRC-97)**, **351 (WRC-03)** and **544 (WRC-03)**, to review the allocations to all services in the HF bands between 4 MHz and 10 MHz, excluding those allocations to services in the frequency range 7 000-7 200 kHz and those bands whose allotment plans are in Appendices **25**, **26** and **27** and whose channelling arrangements are in Appendix **17**, taking account the impact of new modulation techniques, adaptive control techniques and the spectrum requirements for HF broadcasting (Agenda Item 1.13);

– to review the operational procedures and requirements of the Global Maritime Distress and Safety System (GMDSS) and other related provisions of the Radio Regulations, taking into account Resolutions **331 (Rev.WRC-03)** and **342 (Rev.WRC-2000)** and the continued transition to the GMDSS, the experience since its introduction, and the needs of all classes of ships (Agenda Item 1.14);

– to consider the regulatory and operational provisions for Maritime Mobile Service Identities (MMSIs) for equipment other than shipborne mobile equipment, taking into account Resolutions **344 (Rev.WRC-03)** and **353 (WRC-03)** (Agenda Item 1.16).

4 FUTURE OF GMDSS

In Author's opinion, the future of the GMDSS is closely connected with the development of the e-navigation project and with a role of the GMDSS in this process.

For realizing the full potential of e-navigation, the following three fundamental elements should be in place (Korcz, 2009):

1 Electronic Navigation Chart (ENC) coverage of all navigational areas;
2 a robust electronic position-fixing system (with redundancy); and
3 an agreed infrastructure of communications to link ship and shore.

It is envisaged that a data communication network will be one of the most important parts of the e-navigation strategy plan.

In order to realize efficient and effective process of data communication for e-navigation system, the existing radio communication equipment on board (GMDSS), as well as new radio communication systems should be recognized.

The above mentioned GMDSS MF, HF and VHF equipment and systems (Fig. 1) can be also used as a way of data communication for the e-navigation system, provided that this equipment will be technically improved by means of:

– digitization of the analogue communication MF, HF and VHF channels;
– application of high-speed channel to GMDSS;
– utilization of SDR (Software Defined Radio) technology;
– adaptation of IP (Internet Protocol) technology to GMDSS;
– integration of user interface of GMDSS equipment; and
– any other proper technology for GMDSS improvement.

This technical improvement of GMDSS equipment may mean the potential replacement of the conventional equipment by *virtual* one.

In this approach to development of e-navigation it is very important that the integrity of GMDSS must not be jeopardized.

With respect to the radiocommunication aspects required for e-navigation (modernization process), the following should be taken into account as well :

– autonomous acquisition and mode switching;
– common messaging format;
– sufficiently robust;
– adequate security (e.g. encryption);
– sufficient bandwidth (data capacity);
– growth potential;
– automated report generation;
– global coverage (could be achieved with more than one technology).

The next ITU Word Radiocommunication Conference will take place in 2012 (WRC-12) and the main maritime radiocommunication items which will be discussed are following (COMSAR 15, 2011):

– to revise frequencies and channelling arrangements of Appendix **17** to the Radio Regulations, in accordance with Resolution **351**

(Rev.WRC-07), in order to implement new digital technologies for the maritime mobile service; and

- to examine the frequency allocation requirements with regard to operation of safety systems for ships and ports and associated regulatory provisions, in accordance with Resolution **357 (WRC-07)**.

From among other announcements concerning the future GMDSS modification the following should be given:

- the new Arctic NAVAREAs/METAREAs are expected to be transitioned to Full Operational Capability (FOC) on 1 June 2011;
- the new Arctic NAVAREAs/METAREAs are expected to be transitioned to Full Operational Capability (FOC) on 1 June 2011;
- Inmarsat Global Limited (Inmarsat) has informed of its intention to close the Inmarsat-B Service from 31 December 2014;
- the new Cospas-Sarsat MEOSAR system will be probably full operational on 2018.

5 CONCLUSIONS

Twelve years have passed since the time when the Global Maritime Distress and Safety System (GMDSS) became introduced. Planning for the GMDSS started more than 25 years ago, so elements of it have been in place for many years.

There have been numerous advances in the use of maritime radiocommunication to maritime safety, security and environmental protection during this period. But now there are some obsolete GMDSS equipment and systems or the ones that have seldom or never been used in practice.

On the other hand there are a lot of new digital and information technologies.

Not only in the Author's opinion, the time is ripe to start the wide discussion on what the real condition and needs of the marine radiocommunication are, in particular with reference to the current discussion on the e-navigation strategy.

During this work it is necessary first to identify real user needs and secondly to realize that the modernization of the maritime radiocommunication should not be driven by technical requirements. In addition, it is necessary to ensure that man-machine-interface and the human element will be taken into account including the training of personnel.

The lessons learnt from the original development and operation of GMDSS should be taken into account in the modification of GMDSS as well.

Furthermore a systematic process is needed to review and modify the GMDSS to ensure it remains modern and fully responsive to changes in requirements and evolutions of technology and it will meet the expected e-navigation requirements.

For assuming this process a mechanism for continuous evolution of the GMDSS in a systematic way should be created.

At the beginning, for synchronization of this work **a work plan** for the process of the review and modernization of the GMDSS should be established, taking into account the above mentioned issues.

A framework document which defines timescales of this work should be recognized as well.

And finally it should be noted that a key to the success of the review and modernization process is not only that the work is completed on time, but also that it has the flexibility to implement changes ahead of schedule.

In the above context further it should be noted that the Sub-Committee on radiocommunications and search and rescue (COMSAR) since it's last session in 2010 (COMSAR 14) has started work on issue *the Scoping Exercise to establish the need for a review of the elements and procedures of the GMDSS*. Finish of this process is planed on COMSAR 16 in 2012 and it is expected that as the result of this work a lot of answers will be given on the future process of the review and modernization of the GMDSS (COMSAR 14, COMSAR 15).

REFERENCES

Czajkowski J., Bojarski P., Bober R., Jatkiewicz P., Greczycho J., Kaszuba F. & Korcz K. 2000. System GMDSS regulaminy, procedury i obsługa, Wydawnictwo „Skryba" Sp. z o.o., Gdańsk (in Polish)

International Maritime Organization (IMO). 2004. International Convention for the Safety of Life At Sea (SOLAS), London

Sub-Committee on Radiocommunications, Search and Rescue - COMSAR 14. 2010. Report to the Maritime Safety Committee (MSC), International Maritime Organization (IMO), London

Sub-Committee on Radiocommunications, Search and Rescue - COMSAR 15. 2011. Input documents, International Maritime Organization (IMO), London

Korcz K. 2007. GMDSS as a Data Communication Network for E-Navigation. 7th International Navigational Symposium on "Marine Navigation and Safety of Sea Transportation" TransNav 2007, Gdynia

International Telecommunication Union (ITU). Radio Regulations, Geneva 2008

Korcz K. 2009. Some Radiocommunication aspects of e-Navigation. 8th International Navigational Symposium on "Marine Navigation and Safety of Sea Transportation" TransNav 2009, Gdynia

Safety at Sea

Safety at Sea

International Recent Issues about ECDIS, e-Navigation and Safety at Sea – Marine Navigation and Safety of Sea Transportation – Weintrit (ed.)

20. Visual Condition at Sea for the Safety Navigation

M. Furusho
Kobe University, Graduate School of Maritime Sciences, Kobe, Japan

K. Kawamoto
Kawasaki University of Medical Welfare, Okayama, Japan

Y. Yano & K. Sakamoto
Kobe University, Graduate School of Maritime Sciences, Kobe, Japan

ABSTRACT: To the navigation officers of the watch (OOW's) on the navigation bridge the environmental condition of visual acuity is the most important factor for keeping a proper look-out, as regulated by the IMO COLREG's (Rule 5). Navigation officers are required to keep a proper look-out for prevention of (ships) collisions. There are many collision incidents at sea, and especially so under conditions of good visibility. This paper has two topics related to the illuminance and luminance from sunlight, as follows:
The first topic is the introductory explanation of the illuminance inside the navigation bridge.
The second topic is the sky luminance condition as seen by the OOW from the navigation bridge.
These two topics are fundamental factors related to visual perception at sea. The OOW on the bridge has to understand that every care must be taken, especially in fine weather conditions.

1 INTRODUCTION

1.1 *Navigational visual condition at sea*

The meaning of the navigational visual condition at sea is the environmental condition by eye-sight at sea. The visual perception of the target at sea in the maritime traffic system is herein considered, based on the following two measurement results:

1 horizontal illuminance inside the navigation bridge
2 sky luminance of 2 degrees within the horizon

In addition, the visual perception of targets such as aids to navigation and ships is directly related to environmental physical characteristics.

1.2 *A proper look-out*

As you know, Rule 5 of the COLREG's (International Regulations for Preventing Collisions at Sea, 1972) define "Look-out" as follows: Every vessel shall at all times maintain a proper lookout by sight and hearing as well as by all available means appropriate in the prevailing circumstances and conditions so as to make a full appraisal of the situation and of the risk of collision.

1.3 *Marine accidents caused by the improper look-out*

An improper look-out is often pointed out as a cause of marine accidents. Fig. 1 shows the annual change of the total number of marine accidents and the ratio of improper look-out, based on data collected by the MAIA (Marine Accident Inquiry Agency) in Japan, years 2000 ~ 2008.

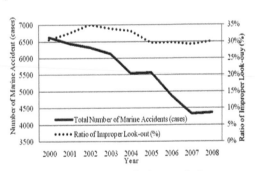

Figure 1. Marine accident caused by improper look-out

2 OBSERVATION

2.1 *Horizontal illuminance in the navigation bridge*

The Captain, navigation officers, look-outs and helmsmen (hereafter termed "navigators") on watch on the navigation bridge are strongly affected by the effect of natural lighting such as sunlight. The illuminance outdoors by day and by night has a wide variance between 100,000 lx from direct sunlight in fine weather and by 0.2 lx from the light of a full moon.

The illuminance meter (model IM-3 made by TOPCON Co. Ltd.) as a measurement device connected with a recording printer was used as indicated in Photo 1. There was no additional condition which had a restriction, such as alteration of the course, speed or others, during this measurement of illuminance on the navigation bridge.

Photo 1, Measurement device

Table 1. Specification of the cooperative ships

Month	Ship	Purpose	G.T. Tons	L.O.A. m	Speed knots	H.E. m
March	C	T	449	49.95	13.5	7
May	A	CF	3,611	114.50	19.5	14
June	B	CF	19,796	192.90	21.8	22
July	C	T	449	49.95	13.5	7
Sept	D	CF	3,597	196.00	25.0	23

Remarks
1) T: Training ship, CF:Car Ferry, H.E.:Height of Eye
2) Observing area was around Japan. Latitude: 35degrees N.

2.2 Sky luminance in 2 degrees from the horizon

One of the conditions to recognise a target, such as a ship or aid to navigation, is the background luminance of the object. It is necessary for visible perception that the difference of the luminance between the background and the target should be more than the value of the luminance difference threshold.

The luminance difference threshold means the threshold limit value of the brightness, based on the experimental studies by Blackwell, H.R., in 1946 and Narisade, K. et al, in 1977 and so on.

About a background when we look at a target, it can be judged from the navigators' characteristics of eye movement at sea. This background is both the sky luminance of 2 degrees above the horizon and sea surface luminance of 2 degrees below the horizon, but in this paper the sky luminance is taken into consideration.

The measurement was carried out on board ship C. The luminance meters (TOPCON's meters BM-5, BM-5A and BM-8: Photo 2) were set and directed right ahead through the windscreen of the navigation bridge according to the regular procedure by navigators. The specification of luminance meter BM-5A is

shown in Table 2. The weather conditions were fine, or fine and cloudy, with direct sunlight.

BM-8   BM-5A

Photo 2. Luminance Meters

Table 2. Specification of the luminance meter(BM-5A)

Optical System	Diameter 32 mm, F=2.5
Measurement Angle	0.1, 0.2, 1, 2 degrees
Photo acceptance Unit	Electronic Light Amplifier
Wave Length	380-780 n.m.
Range	$0.0001 \sim 1200000 cd/m2$
Distance	$520mm \sim \infty$
Sampling Time	2 sec
Size & Weight	355(L) X 130(W) X 169(H) mm , 4 Kg

3 RESULTS

3.1 Horizontal illuminance on the navigation bridge

The result of the measurement data of the horizontal illuminance on the navigation bridge using the illuminance meter connected with a printer, as indicated in Photo 1, is shown in Fig. 2. The vertical scale in Fig. 2 shows illuminance in lux in logarithmic scale; the horizontal scale shows Japan Standard Time (JST) which has 9 hours difference from Greenwich Mean Time (GMT). These data have dispersion which depends on the different observing times in a month.

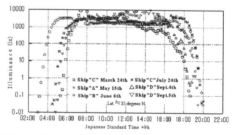

Figure 2. Horizontal illuminance in the navigation bridge

3.2 Sky luminance of 2 degrees above the horizon

The sky luminance, dependent on the relative direction towards the solar direction in the open sea, was measured. The result of these observations is shown in Table 3. These experimental observations were carried out in July during a research voyage of the training ship.

Table 3 Sky luminance of the relative direction towards the solar direction at open sea

Relative Angle degrees	Max.:A cd/m²	Min.:B cd/m²	Range:A-B cd/m²
0	27,420	7,070	20,350
45	20,900	6,610	14,290
90	6,627	4,723	1,904
135	5,500	3,960	1,540
180	6,165	4,166	1,999
-135	6,095	3,910	2,185
-90	6,170	4,249	1,921
-45	12,190	6,340	5,850

Remarks: Relative angle 0 degree means the solar direction.
"-(negative number)" means the left side of solar direction

The sky luminance, dependent on the relative direction towards the solar direction at the time of relative angle around 0 (zero) degrees, changes from 7,000 to 27,000 cd/m². The range, which means the difference of luminance between the maximum value at A and the minimum value at B, was found to be approximately 20,000 cd/m².

The opposite side, in the case where the relative angle is -90 ~ +90 degrees of solar direction, produced a small change at 1/10th of the range.

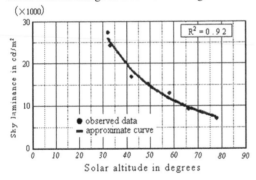

Figure 3. Relationship between the sky luminance and the solar altitude in degrees.

4 CONSIDERATION

4.1 Horizontal illuminance by the standardization with using solar altitude

Fig. 2 shows the seasonal difference for times of sunrise and sunset; also, the difference in hours of morning or evening twilight at differing sea areas.

There is no affect by the different heights of eye on the horizontal illuminance on the navigation

bridge. The illuminance change in daytime has a wide value between 1,000 lx and 10,000 lx, but at morning or evening twilight the illuminance changes rapidly with time. This is a remarkable feature of twilight – at this time there are functional changes of both the cone and the rod of the visual cell.

The horizontal illuminance nearby the windshield on the navigation bridge has various changes between 0.01 lx and 10,000 lx, according to the voyage situation such as seagoing area, navigation time, ship's course and so on.

Because the illuminance change has seasonal characteristics according to the times of sunrise, sunset and the hours of twilight, standardisation of illuminance based on the observation time might be difficult. Therefore, the solar altitude is useful for the standardisation – by using calculated solar altitudes based on the observing time and geographical position.

Fig.4 shows the results of the standardisation by using solar altitude. This demonstrates that the solar altitude is a suitable factor for explaining the change of horizontal illuminance on the navigation bridge.

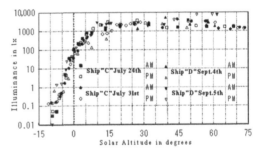

Figure 4 Horizontal illuminance on the navigation bridge by the standardisation method using solar altitudes

4.2 Sky luminance on shore

For the purpose of comparing the luminance at sea and ashore, the example of illuminance measurement at Fukui prefecture in Japan is taken into account (Ref. Lighting Handbook published by the Illuminating Engineering Institute of Japan)

Fig.5 shows this example which had no observation data under 5 degrees on shore. In this figure, the line on the celestial sphere via the sun and the (observer's) zenith is characterised as bilaterally symmetric, the maximum point being marked as " X " near the sun and the minimum point, 90 degrees distant via the zenith, as " ● ".

4.3 Sky luminance at sea in fine weather

The sky luminance in the relative solar direction is shown in Fig.6 according to the solar altitude. The numbers indicated around this radar charted figure show the relative angle in degrees and the numbers indicated on the radial axes show the sky luminance in cd/m^2.

The distribution of sky luminance to each relative solar direction in fine weather (with sunlight) gradually becomes concentric circles. When the solar altitude is more than 60 degrees the variability of the luminance is small, so the background condition of the visual perception is uniform.

Comparing with the case on shore (Fig. 5), it is understandable that items have similar conditions, as follows:

1 There is a maximum point of the sky luminance relative to the solar direction.
2 There is a minimum point in the opposite direction to the maximum point (when the relative solar direction is near 180 degrees).
3 The line on the celestial sphere via the sun and the zenith is characterised as being bilaterally symmetric.
4 The distribution of the sky luminance has uniformity with no relationship to the relative solar direction, nor to the solar altitude.

These items are the evidence to support the navigators' statements which explain that it is easy to recognise targets visually when "the sun is behind me, not in front of me".

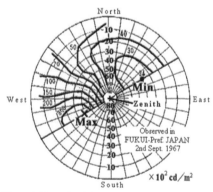

Figure 5 Example of the sky luminance on shore

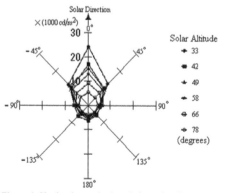

Figure 6 Sky luminance in the relative solar direction

4.4 Sky luminance of the solar direction

The approximate curve as shown in Fig.3 is taken into consideration. The fractional approximate curve of the sky luminance of the solar direction related to solar altitude can be explained by the formula as shown hereunder (1).

$$Y = a/X + B \qquad (1)$$

where, Y=sky luminance in cd/m^2; X=solar altitude in degrees; a=105 (coefficient); b= -5,000 (constant). According to Fig.4, this formula can be applied when the solar altitude is more than 10 degrees.

4.5 Relationship between the solar altitude and ship's collision

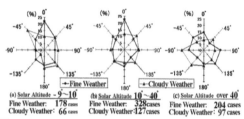

Figure 7 Impact by the Solar altitude and the relative direction on the ships' collision

According to the result of analysing 1000 cases((a)244, (b)455, (c)301) of ships' collisions, Fig. 7 can be obtained as an impact by the solar altitude and the relative direction toward the solar direction.

The solar direction in the case of less than 40 degrees of the solar altitude has a direct effect to the ships' collision.

The OOW should remind not only the solar direction but also fine weather. Generally speaking, we believe that the fine weather is good weather, but fine weather might have the blind spot so-called "white hall" which means the restricted visual condition for the proper look-out by the OOW at sea.

5 CONCLUSIONS

The navigational visual, environmental background condition of recognising targets (so-called 'visual perception') by sight at sea has been taken into consideration.

There are 4 conditions which should be considered, as shown hereunder, in order to discuss the visual perception:

1 Luminance of a target
2 Luminance of the background
3 Adaptation condition of the (observer's) retina
4 Equivalent Veiling Luminance from the near visual field

In this study the authors have focused on the above item 2) – luminance of the background – because of the necessity for the field of study on board ship.

5.1 *Illuminance on the navigation bridge*

The horizontal illuminance on the navigation bridge can be standardised based on the solar altitude, as shown below in table 4:

We can understand the illuminance condition of foreside on the navigation bridge. The start and end of navigational twilight is -9(minus nine) degrees.

Table 4 Horizontal illuminance on the navigation bridge

Solar altitude degrees	Horizontal illuminance lx
over 10	1000 ~10,000
10	1000
0	100
-3	10
-6	1
-9	0.01 ~ 0.1 *
-18	Start and end of astronomical twilight

Note * The horizon cannot be seen except in the solar direction at the start and end of navigational twilight.

5.2 *Luminance condition for the OOW*

The sky luminance at 2 degrees above (and below) the horizon is one of the most important composition factors of visual perception at sea.

The authors obtained the remarkable features on the sky luminance based mainly on the experimental observations on board, and these are shown as follows:

1 The value of the sky luminance at 2 degrees above the horizon is bigger than the sea surface luminance, except in cases of sun-glitter on the sea surface
2 The sky luminance of the solar direction has various changes between approx. 7,000 and 27,000 cd/m^2
3 The sky luminance of the solar direction related to the solar altitude can be explained when the solar altitude is more than 10 degrees by this formula:

$Y = a/X + B$

where, Y=sky luminance in cd/m2; X=solar altitude in degrees; a=105 (coefficient); b= -5,000 (constant).

4 The distribution of sky luminance to each relative solar direction gradually forms concentric circles. When the solar altitude is more than 60 degrees the variability of the luminance distribution is small and the background condition of the visual perception can be said to be uniform
5 The distribution of the sky luminance on the opposite side of the solar direction has uniformity, with no relationship to the relative solar direction nor to the solar altitude. This is the evidence to support navigators' statements which say that it is "easy to recognise targets when the sun is behind the observer".
6 The OOW should think of the "white hall" which means gimmick of the proper look-out especially in fine weather.

ACKNOWLEDGMENT

This work was supported by the Grant-in-Aid for Scientific Research (B) (KAKENHI) No21300211.

REFERENCES

Narisada, K Yoshimura,Y. (1977): Adaptation luminance of driver's eye at the Entrance of Tunnel ---an Objective Measuring Method, Transactions of the 3rd International Symposium of Road Lighting Effectiveness, Karlsruhe, On pp. 5-6.

Lighting handbook ,The Illuminating Engineering Institute of Japan (IEIJ).

Vos, J. J. (1984), Disability glare - a state of the art report, CIE Journal, Vol.3, No.2.

Narisada, K. (1992), Visual Perception in Non-uniform fields, Journal of Light & Visual Environment, Vol.16, No.2.

Minnaert, M.G.J., Translated and Riviced by Len Seymour (1993), Light and color in the outdoors, p.102 - p.125, Springer-Verlag.

Furusho, M. (1995), Visual Perception of Horizon for a Good Lookout at Sea, The Journal of Japan Institute of Navigation, Vol.93, pp.35-42, In Japanese.

Furusho, M. (1997), Visual Environment for a Good Lookout at Sea, The Journal of Japan Institute of Navigation, Vol.96, pp.79-86, In Japanese

Furusho, M., Machida, K., Fujioka, Y. (1998), A Study of a Good Lookout at Sea in Case of Ship Collisions, The Journal of Japan Institute of Navigation, Vol.99, pp.101-106, In Japanese

Furusho, M. (1999), A Study of Visual Perception of the Ships for a Good Lookout at Sea, The Journal of Japan Institute of Navigation, Vol.100, pp.59-66, In Japanese

21. Safety Control of Maritime Traffic Near by Offshore in Time

D. Yoon
SUNY Maritime College and Mokpo National Maritime University

M. Yi, J. S. Jeong & G. K. Park
Mokpo National Maritime University

N. S. Son
MOERI, Korea Ocean Research and Development Institute

ABSTRACT: This paper introduces and presents a strategy for ensuring safety during control of vessel interaction in real time. A measure of danger during the interaction is explicitly computed, based on factors affecting the impact force during a potential collision between an object and the vessel. A motion strategy to minimize the danger factors and risk level is developed for articulated degree of freedom for multi activities of vessel. Simulations and experiments demonstrate the efficacy of this approach. The aim of this research is to introduce and develop an assistant system by analyzing ships activities in real time from land aspect.

1 INTRODUCTION

As vessel move from the water point to water point and other water environments, safety of safety controls interacting with vessel becomes a key issue [1, 2]. The vessel system must provide a mechanism to ensure ship safety, given the uncertain environment and untrained sailors. To ensure the safety and intuitiveness of the interaction, the complete system must incorporate (i) safe mechanical design, (ii) safety control friendly interfaces such as natural language interaction and (iii) safe planning and control strategies. Our work focuses on this third item. In particular, the goal is to develop strategies to ensure that unsafe sailing does not occur between any point on an articulated vessel and a safety control in the vessel's workspace. This paper focuses specifically on the real time safety during safety control of vessel interaction. The concept of e-navigation can be drawn as below figure 1. The process of e-navigation will affect whole process of development. The process of practice and operation is affected by user requirement, operational function, and technical equipment and gear.

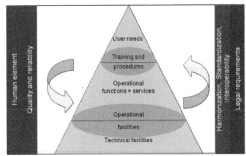

Figure 1. IMO Strategy for E-Navigation Process

2 DESIGN A MODEL FOR SAFETY CONTROL IN REAL TIME

In this research, there are several navigation equipment were used to analyze navigation data including from ARPA, AIS, NAVTEX and VHF. It is assumed that those data were received the data field from low data given by each navigation aids. In brief, the information from ARPA and AIS is as follows,
– Present bearing of the target
– Present range of the target
– Predicted target range at the closest point of approach (CPA)
– Predicted time to CPA (TCPA)
– Calculated true course of the target
– Calculated true speed of the target

- IMO number, Call Sign, Name, Length, Beam, Type of ship, Location of position-fixing antenna on the ship
- Ship's Position, Time in UTC, Course over ground, Speed over ground, Heading, Navigational status, Rate of turn, Ship's draught, Hazardous cargo, Destination and ETA

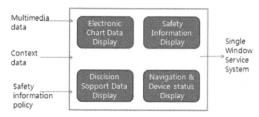

Figure 2. Integrated Service System Using AIS

Based on that information, this paper applied a model of meaning analysis which is made by each data field by navigation gear and target-related knowledge. The model can show just simple sentence. In next, by using the meaning analysis for designed navigational equipment, it is showed a model of meaning model for target.

Case 1: If target is ship or vessel,
- Usually by performing ARPA, AIS, and VHF, the information can be obtained. Therefore, in this case, the model of meaning analysis is combined the information of ARPA, AIS, and VHF. Below figure can explain this case.
-

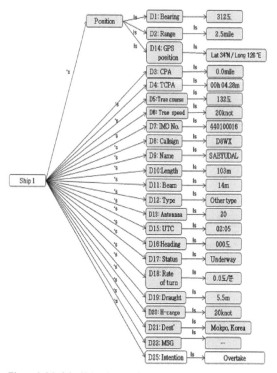

Figure 3. Model of Meaning Analysis: If target is Vessel
1

Case 2: If target is other than ship or vessel,
- Usually in this case, the model of meaning analysis is combined the information of ARPA and NAVTEX because it is impossible to get information from AIS and VHF. Below figure can explain this case.

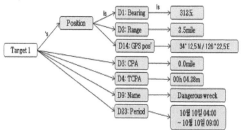

Figure 4. Model of Meaning Analysis: If target is Other Than Vessel
1

The view of intelligent safety information system and discrete event system
- Step of unit filtering and recognition: need to analyze data coming up in real time.
- Step of expecting situation: need to check out in terms of discrete event system.

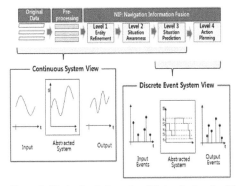

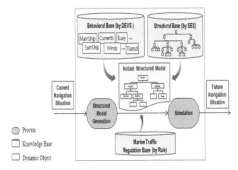

Figure 5. Navigation Information Mixing System by Discrete Event System View

– System Draft for Expecting Safety Situation

Figure 6. Draft of Simulation Module for Expecting Situation

1 Three Dynamic Units
   – Input Factors: Current Navigation Situation
   – Output Factors: Future Navigation Situation
   – Inside Unit: Instant Structured Model for Dynamic Change of Situation

Identification Model for Degree of Collision Risk

2 Three Knowledge Base including Behavioral Base, Structural Base (Behavioral Base SES(System Entity Structure)), Regulation Base

Two Inside Process for Structured Model Generation and Simulation Process

Blocking area theory is effective to avoid single traffic ship but it has difficulty in avoiding lots of traffics concurrently in the real sea. Through simulator experiments, it can be found that collision risk is estimated normally by using fuzzy algorithm, with the similar tendency of environmental stress in open sea and confined waterway. It can be drawn in figure 7.

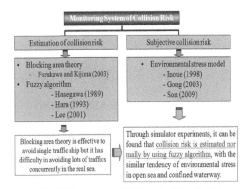

Figure7. Model Research for Risk Level of Ship Collision

Figure 8 showed a flow of an assessment algorism of degree of collision risk which is used for safe ship movement planning and control. Based on marine traffic data, interrelated model can be summarized and described in below Figure 9.

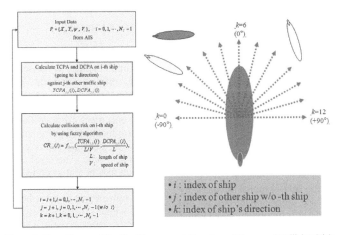

Figure 8. Flow Chart and Draft of Assessment Algorism of Degree of Collision Risk

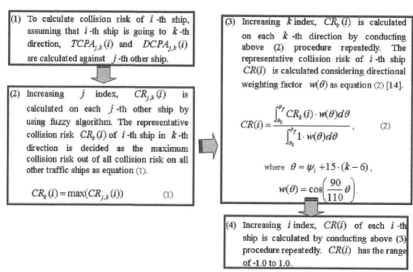

(1) To calculate collision risk of i-th ship, assuming that i-th ship is going to k-th direction, $TCPA_{j,k}(i)$ and $DCPA_{j,k}(i)$ are calculated against j-th other ship.

(2) Increasing j index, $CR_{j,k}(i)$ is calculated on each j-th other ship by using fuzzy algorithm. The representative collision risk $CR_k(i)$ of i-th ship in k-th direction is decided as the maximum collision risk out of all collision risk on all other traffic ships as equation (1).

$$CR_k(i) = \max(CR_{j,k}(i)) \qquad (1)$$

(3) Increasing k index, $CR_k(i)$ is calculated on each k-th direction by conducting above (2) procedure repeatedly. The representative collision risk of i-th ship $CR(i)$ is calculated considering directional weighting factor $w(\theta)$ as equation (2) [14].

$$CR(i) = \frac{\int_{\theta_0}^{\theta_f} CR_k(i) \cdot w(\theta) d\theta}{\int_{\theta_0}^{\theta_f} 1 \cdot w(\theta) d\theta}, \qquad (2)$$

where $\theta = \psi_i + 15 \cdot (k - 6)$,

$w(\theta) = \cos\left(\frac{90}{110} \theta\right)$.

(4) Increasing i index, $CR(i)$ of each i-th ship is calculated by conducting above (3) procedure repeatedly. $CR(i)$ has the range of -1.0 to 1.0.

Figure 9. Chart of Assessment Algorism of Degree of Collision Risk

3 TEST RESULT FOR RISK ASSESSMENT

This research showed the test result of whole concept models in order to verify the assessment model of risk degree of environment near the vessel. The structure of test will be shown in the below figure. Dotted line is to be designed. Experimental frame's Generator provides random environmental data for verify outputting data. In the step of Pre-processing, data will be treated and processed in order to be understandable after receiving data. After then, input data will be used to calculate the assessment result of risk degree for each parameter after passing fuzzy professional system. Finally, Total_ERAN Process will guide and explain total degree of collision risk and unit risk degree to the user and/or sailor.

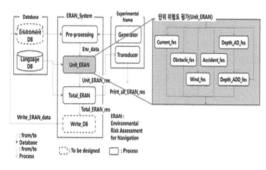

Figure 10. Chart of Test System

The experimental result is showed implementation for random situation and assumed scenario by generator. In the data base, environment DB are important parts in real time information and containing pre-decided environment risk list in its activating

equipment. however in this research, real time environment information are created by generator which are investigating case for conception model , except in here Result from environment risk assessment by random cases from generator, Maritime safety information which is connected and unified to GI-COMS (General Information Center on Maritime Safety & Security) makes it possible to revaluate its effectiveness and reproduce

- improve governmental, people require service .By provide suitable information which are necessary to government division and company
- improve safety for crew, cargo, vessel by providing pre-alarm in accident frequent site and danger site of pirate
- protect personal and national information by internet based on GICOMS operation security system

4 CONCLUSIONS

This study integrates various theories and methodologies implemented in vessel safety systems into a unique, original platform. Ultimate goal was the complete integration of a vessel safety, which in turn is able to promote the safety, security and comfort of vessel occupants from collision. Therefore, this paper introduced and presented a model for ensuring ship safety during a safety control of vessel interaction in real time. The level of danger in the interaction due to a potential collision is explicitly defined as the danger characteristic. A sequential onestep ahead trajectory planner (the safety system) is presented which generates vessel motion by minimizing the danger characteristic. The algorithm can be used for redundant or non-redundant manipulators, and operates correctly at all vessel configurations.

REFERENCES

[1] A. Bicchi, S. L. Rizzini, and G. Tonietti, "Compliant design for intrinsic safety: General Issues and Preliminary Design," presented at IEEE/RSJ Int. Conf. on Intelligent Vessel and Systems, pp. 1864-1869, 2001

[2] A. J. Bearveldt, "Cooperation between Man and Vessel: Interface and Safety," presented at IEEE Int. Workshop on Vessel Human Communication, pp. 183-187, 1993.

[3] B. Martinez-Salvador, A. P. del Pobil, and M. Perez-Francisco, "A Hierarchy of Detail for Fast Collision Detection," presented at IEEE/RSJ Int. Conf. on Intelligent Vessel and Systems, pp. 745-750, 2000.

[4] N.S. Son, S.Y. Kim, J.Y. Oh, "STUDY ON AN ALGORITHM FOR THE ESTIMATION OF COLLISION RISK AMONG SHIPS BY USING AIS DATABASE", Proceedings of 9th Asian Conference on Marine Simulator and Simulation Research, pp.81-87, 2009

[5] J. Zurada, A. L. Wright, and J. H. Graham, "A Neuro-Fuzzy Approach for Vessel System Safety," IEEE Transactions on Systems, Man and Cybernetics - Part C: Applications and Reviews, vol. 31, pp. 49-64, 2001.

[6] J. Y. Lew, Y. T. Jou, and H. Pasic, "Interactive Control of Human/Vessel Sharing Same Workspace," presented at IEEE/RSJ Int. Conf. on Intelligent Vessel and Systems, pp. 535-539, 2000.

[7] N. S. Son, et. al., "Study on the Collision Avoidance Algorithm against Multiple Traffic Ships Using Changeable Action Space Searching Method,"Journal of Korean Society for Marine Environmental Engineering, Vol.12, No.1, pp.15-22, February 2009.

[8] Mira Yi, Gyei-Kark Park, and Jongmyeon Jeong, "DEVS Approach for Navigation Safety Information System", International Conference on Electronics, Information, and Communication, 2010

[9] O. Khatib, "Real-Time Obstacle Avoidance for Manipulators and Mobile Vessel," The Int. Journal of Vesselics Research, vol. 5, pp. 90-98, 1986.

[10] P. I. Corke, "Safety of advanced vessel in human environments,"Discussion Paper for IARP, 1999.

[11] "RIA/ANSI R15.06 - 1999 American National Standard for Industrial Vessel and Vessel Systems - Safety Requirements." New York: American National Standards Institute, 1999.

[12] S. P. Gaskill and S. R. G. Went, "Safety Issues in Modern Applications of Vessel," Reliability Engineering and System Safety, vol. 52, pp. 301-307, 1996.

[13] Y. Yamada, Y. Hirawawa, S. Huang, Y. Umetani, and K. Suita, "Human - Vessel Contact in the Safeguarding Space," IEEE/ASME Transactions on Mechatronics, vol. 2, pp. 230-236, 1997.

[14] Y. Yamada, T. Yamamoto, T. Morizono, and Y. Umetani, "FTABased Issues on Securing Human Safety in a Human/Vessel Coexistance System," presented at IEEE Systems, Man and Cybernetics SMC'99, pp. 1068-1063, 1999.

[15] Y. Yamada, Y. Hirawawa, S. Huang, Y. Umetani, and K. Suita, "Human - Vessel Contact in the Safeguarding Space," IEEE/ASME Transactions on Mechatronics, vol. 2, pp. 230-236, 1997.

[16] V. J. Traver, A. P. del Pobil, and M. Perez-Francisco, "Making Service Vessel Human-Safe," presented at IEEE/RSJ Int. Conf. on Intelligent Vessel and Systems (IROS 2000), pp. 696-701, 2000.

22. Maritime Safety in the Strait of Gibraltar. Taxonomy and Evolution of Emergencies Rate in 2000-2004 period

J. Walliser, F. Piniella, J.C. Rasero & N. Endrina
Departamento de Ciencias y Técnicas de la Navegación. Universidad de Cádiz. CASEM – Facultad de Ciencias Náuticas, Puerto Real, Cádiz, Spain

ABSTRACT: Both SAR'79 and UNCLOS'82 Conventions are specific tools that establish the juridical and technical foundations for the development of reactive aspects related to maritime safety response. These conventions set up the search and rescue regions in which coastal states should assume the responsibility to dedicate resources, to cover the needs of the SAR responsibilities. 2006 amendments to IAMSAR manual volume I, in force since 2007, June the 1st, established the identification and assessment of risks related to maritime safety as one of the practical principles in maritime risk management. The Strait of Gibraltar is a narrow navigational channel connecting the Atlantic Ocean and the Mediterranean Sea between Spain and Morocco. The Strait supports a huge volume of maritime traffic increasing steadily every year. This paper presents the preliminary results obtained in relation with the taxonomy and temporal distribution of maritime emergencies reported and documented by the Spanish Maritime Administration throughout 2000-2004 period.

1 BACKGROUND

The active involvement of the coastal states in safeguarding and promoting the safety of human life, environment and property related to maritime navigation in the waters in which they exercise jurisdiction, sovereignty or sovereign rights, is shown in a number of international legal texts.

The International Convention on Maritime Search and Rescue, Hamburg, 1979 (Convention SAR'79) and the United Nations Convention for the Law of the Sea, Montego Bay, Jamaica, 1982 are included among those juridical tools.

These two fundamental legal instruments lay down both regulatory and technical aspects of the development of reactive response to maritime emergencies.

Both texts establish the principle of division of the entire maritime waters, defining areas of responsibility for maritime search and rescue associated with every coastal nation. These nations should assign specific resources - human, technical and legal - to meet the requirements that arise as a result of the liabilities undertaken by the parties.

Although both conventions regulate the commitments related to maritime search and rescue matters undertaken by the parties, the International Convention on Maritime Search and Rescue, Hamburg 1979, known as SAR'79 Convention, which Spain joined in 1993, lays down the basic

guidelines to be followed by the Authorities of the coastal states in the process of design and implementation of maritime search and rescue services.

Over the years, this agreement has been amended a number of times. Among the amendments which are due to be highlighted, we find those adopted in 1998. According to these, it is essential to provide the centers responsible for carrying out maritime search and rescue operation with detailed operational plans appropriate and adapted to the particularities of each specific search and rescue region. These plans will allow carry out these actions effectively.

These plans should also establish not only the procedures to be followed during mobilization of rescue units, but also provide the methodology to be used in developing search and rescue operations. The plans require the establishment of coordination instruments between adjacent rescue centers and procedures and criteria to be used not only during the gathering and evaluation of relevant information related to the emergency but also alerting ships and aircraft transiting the area of the incident and requesting their cooperation in operations.

A rigorous approach to the formulation and development of plans and protocols to cope with maritime emergency situations requires the application of specific methodological tools that would make possible the identification, classification and categorization of those risks that must be controlled or the mitigation of their consequences.

On the other hand, at the request of the interested countries, and with the aim of facilitating the adoption of the necessary measures for the adaptation of the standard and promoting harmonization in a global environment, the International Maritime Organization (IMO) and International Civil Aviation Organization (ICAO) published in 1999, the International Aeronautical and Maritime Search and Rescue Manual (IAMSAR), a dynamic document that over the years has undergone several modifications which enabled it to improve and adapt to changing reality.

Among these changes we can find the 2006 amendments, in force since June 1, 2007, which set out the practical principles to be followed in the implementation of aeronautical and maritime search and rescue services and addresses the need to identify and asset the risks related to maritime safety. In such a way that these amendments state that the effectiveness of the response to maritime emergencies depends, among others, upon the knowledge of type and frequency of those marine incidents that may result in a risk for human life at sea, safety of navigation and protection of marine and coastal environment.

This paper presents the preliminary results of the study carried out over the emergency rates in the Strait of Gibraltar, approaching the taxonomy and distribution of incidents and accidents documented by the Spanish Maritime Administration from 2000 to 2004 in the geographical area of the Strait of Gibraltar.

2 ENVIRONMENT

2.1 *Geographical environment*

The Strait of Gibraltar (Figure 1) is the natural passage which links the Mediterranean Sea with the Atlantic Ocean. Although its boundaries have never been formally established, for the present research study the western boundary has been defined by the line connecting Cape Trafalgar with Cape Spartel while the eastern one has been considered by the opposition Europe Point - Punta Almina.

The European coast, limited by Cape Trafalgar and Europe Point, is 55 nautical miles long, whereas the African coast from Cape Spartel to Punta Almina is 42 nautical miles long.

Its longitudinal axis is divided into two sections. The Western section, some 18 nautical miles in length and oriented approximately east - west, runs from the line connecting the Island of Tarifa, located on the Spanish coast, with Cala Grande, on the African coast, towards the Atlantic Ocean. The

Eastern section runs from west by southwest to east by northeast, along some 15 nautical miles to reach the eastern Strait boundary.

The channel presents its maximum width, 24.2 nautical miles, on its western limit, between Cape Trafalgar and Cape Spartel, while the narrowest section is defined by the line connecting median point between Tarifa and Punta Guadalmesi River, on the northern coast, and Punta Cires, on the southern coast. At this point, the channel is 7.45 nautical miles wide. The eastern embouchure has a maximum width of 12.5 nautical miles.

On the northern coast there are significant shoals and reef areas alternating with broad bays and sandy beaches. On this coast are located the ports of Algeciras- La Línea, Tarifa and Gibraltar.

The southern coast, geologically very similar to the northern one, has, however, a much more rugged and inaccessible coastline. On this coast we find the ports of Tangier, located east of Punta Malabata, Tangiers – Mediterranean, close to Punta Cires, and Ceuta, located by Punta Almina, on the eastern end of the African coast.

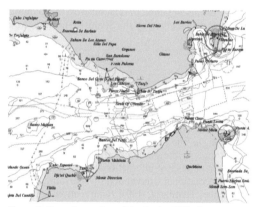

Figure 1. General overview of the Strait of Gibraltar

2.2 *Traffic patterns*

The maritime traffic through the Strait of Gibraltar follows a clearly defined pattern which is conditioned by four basic parameters: the last port of call of the ship, destination port, the routeing measures established and prevailing weather conditions.

In general the flow of maritime traffic follows two fundamental axes (Figure 2). The most important in terms of traffic density, is the longitudinal axis defined by the tracks of the ships passing from the Mediterranean Sea towards the Atlantic Ocean and vice versa.

The second axis is defined by the tracks of the vessels, mainly ferry ships and High Speed Crafts, connecting the ports located on both sides of the Strait.

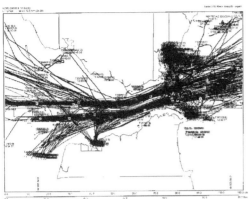

Figure 2. Cumulative maritime traffic radar surveillance picture
Source: Tarifa VTS

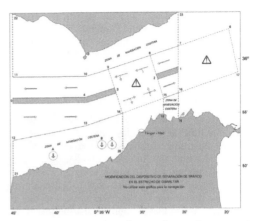

Figure 4. Traffic separation scheme and precautionary areas in the Strait of Gibraltar Source: IMO.

The combination of a very high traffic density area (94,157 transits identified through year 2005) (Figure 3), the existence of high concentration of crossing tracks and occasionally very unfavorable weather conditions within a narrow channel, have forced the Spanish and Moroccan governments to promote, through the International Maritime Organization, the establishment of several marine traffic organization and monitoring measures (Figure 4) – traffic separation scheme, mandatory reporting system, precautionary areas, vessel traffic services – all of them complemented by an extensive network of maritime signals covering both, northern and southern coasts.

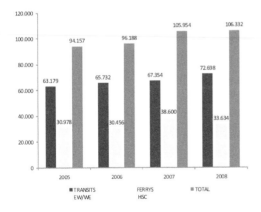

Figure 3. Traffic evolution. Years 2005-2008
Source: Tarifa VTS.

3 HISTORICAL ANALYSIS OF MARITIME EMERGENCIES IN THE STRAIT OF GIBRALTAR

3.1 *Documentary sources and research structure*

The historical analysis of emergencies in the Straits has been developed on the basis of the information provided by three historical data sets:
1 The database of the General Directorate of Merchant Marine.
2 Annual reports of the Maritime Rescue Coordination Centre of Tarifa, and
3 Annual reports of the Maritime Rescue Coordination Centre of Algeciras.

The analysis of the three data sets reveals a lack of harmonization that, although the big similarities in their basic structure, leads to significant differences in the criteria followed while classifying the information related to each event. It it should be highlighted the lack of information concerning the initiating events or causes of accidents and the extent of damage to property.

Nevertheless, despite the lack of some important information, the geographical location of each event and a detailed description of the units deployed during the response operations are always available.

Due to the high level of dispersion of information regarding maritime emergencies occurred in the Strait of Gibraltar and the lack of clearly defined relationships among the historical records, it becomes necessary a previous compilation of the information, and the unification of criteria for the classification and its subsequent analysis. This all led to the creation of a special data base called GIBSAR, which is the basic tool for the development of this historical analysis of maritime accidents in the Strait of Gibraltar.

This database does not only compile information provided by the marine rescue coordination centers of Algeciras and Tarifa and the statistical series of data produce by the General Directorate of Merchant Marine, but it also establishes a scale for the assessment of effect, both for individuals an vessels, caused by the consequences of these events.

3.2 *Analysis of maritime emergencies. Global data*

The analysis of the data (Table 1) shows that the total maritime emergencies documented by the Spanish Maritime Administration in the period 2000-2004 within the Strait of Gibraltar area comes to a total amount of 1,216 cases. Only 23.5% of this these figures, 284 cases, are due to false alarms while the remaining 76.5%, 922 cases, are due to real alerts. Notice that the annual distribution of the data presents a great homogeneity.

Table 1. Distribution of real and false alerts related to maritime emergencies in the Strait of Gibraltar. Years 2000-2004

Emergencies	Year					
	2000	2001	2002	2003	2004	Total
Real	181	187	181	189	184	922
False	61	66	51	58	48	284
Total	242	253	232	247	232	1206

Table 2 shows that 90.8% of total number of real cases are related to incidents, that means there is neither major structural damages to ships nor losses of human lives nor missing persons nor pollution episodes involved in the event.

On other hand, only 9.2% of the real emergencies are related to marine accidents, considering such events as those involving total loss of the ship or major structural damages, or losses human lives or missing persons or a pollution episode.

In this case, the distribution is also very homogeneous, with values ranging from a minimum of 181 emergencies documented in 2000 and 2002 to a maximum of 189 emergencies in 2003.

Table 2. Distribution of accidents and incidents related to real maritime emergencies in the Strait of Gibraltar. Years 2000-2004

Emergencies	Year					
	2000	2001	2002	2003	2004	Total
Accidents	20	19	14	12	20	85
Incidents	161	168	167	177	164	837
Total	181	187	181	189	184	922

Table 3 shows the distribution of all real emergencies attended in the period 2000 to 2004 according to subtype.

There is a clear predominance with 24.1% of the total amount of performed search and rescue operations due to mechanical failure of ships systems and/or services. Pollution episodes amounts 15.6% of the total number of cases.

We should also focus the attention on the number of search and rescue operations related to illegal immigration, whether in the preventive stage, escorting the crafts used in the passage through the Strait as well as in the phase of assistance to the occupants or, where appropriate, during search, location and recovery of corpses. These cases come to represent 12.8% of search and rescue operations.

Operations related to drifting objects varies from 5% to 10%, assistance to users of recreational crafts and devices amount 8.4% and 7.2% respectively, rescue operations on coast and cliffs 5.6%, and drifting boats, which comes to be 5% of the total.

We should also consider the medical transfers between Spanish hospitals located on both sides of the Strait (Ceuta to Seville or Cadiz), which means 4.6% of total operations, nearly twice the rate of medical evacuations from ships which rates 2.6% of the total.

Table 3. Distribution of real emergencies according to subtype. Strait of Gibraltar. Years 2000-2004

Subtype	Year					
	2000	2001	2002	2003	2004	Total
Leisure crafts	12	11	8	22	13	66
Assistance to navigation	2	5	3	3	6	19
Overdue	0	3	0	1	1	5
Allision / Collision	2	3	3	0	2	10
Pollution	29	35	35	23	22	144
Drifting crafts	12	7	7	11	9	46
List / Stability	0	0	1	0	0	1
Medical Evacuation	5	9	5	4	1	24
Medical Transfer	4	3	11	7	17	42
Mechanical Failure	37	38	40	50	57	222
Man Overboard	5	4	3	3	4	19
Sinking	2	3	4	3	4	16
Fire/ Explosion	2	4	2		2	10
Illegal Immigration	23	23	19	38	15	118
Drifting Objects	18	18	24	8	9	77
SOS Message	1	0	0	2	0	3
Castaway Rescue	0	0	0	1	1	2
Coastal Rescue	17	13	8	4	10	52
Grounding	3	3	3	4	6	19
Leaking	4	2		2	2	10
Other	3	3	5	3	3	17
Total	181	187	181	189	184	922

Regarding the severity of the consequences on persons, crafts or vessels and on the environment, it should be notice that minor and negligible severity cases range nearly 30%, while moderate severity cases range 30.4% and major severity 28.5%. Severe cases involving total losses of ships, losses of human lives, missing persons or severe pollution events range 6.5% (Table 4).

Table 4. Distribution of real emergencies according to severity rate. Strait of Gibraltar. Years 2000-2004

Severity	Year					
	2000	2001	2002	2003	2004	Total
Severe	15	13	12	9	11	60
Major	50	61	54	58	40	263
Moderate	57	49	47	77	84	314
Minor	14	17	8	8	9	56
Negligible	45	47	60	37	40	229
Total	181	187	181	189	184	922

According to table 5 and regarding casualty condition, almost 90% of the persons involved were assisted or rescued, while 6% of the total amount of persons involved in emergencies got safe by their own means and the number of persons who lost their lives or were missing rates 2.2%.

Table 5. Distribution of real emergencies according to casualty condition. Strait of Gibraltar. Years 2000-2004

Casualty Condition	Year					
	2000	2001	2002	2003	2004	Total
"Shelf rescued"	328	34	6	12	2	382
Rescued	431	825	474	2156	542	4428
Assisted	9	398	373	303	198	1281
Evacuated	65	11	16	12	17	121
Died before arrival	10	27	17	2	1	57
Died after arrival	22	4	12	12	9	59
Missing	5	3	7	9	1	25

4 CONCLUSION

Annual distribution of emergencies, considering the whole period, presents a very stable trend, ranging from a minimum value of 181 to a maximum of 189 cases.

In terms of geographical distribution, two main areas support the highest rate of emergencies both quantitatively and qualitatively. Those areas are the central zone of the Strait of Gibraltar and Algeciras Bay (Figures 5 and 6).

The results of the study highlighted the large number events, such as medical transfers, pollution incidents in port service waters, operations related to the use of recreational crafts and devices, swimmers, diving and other natures related events which, although not considered as maritime emergencies, required the deployment of specific marine search and rescue resources.

It should be noticed the low rate of emergencies directly related to any maritime search and rescue service responsibilities such as leakage, collision or allision, fire or explosion, heel or stranding, which all together rate 7.2% of the SAR operations performed. This rate would increase up to 9.8% if

medical evacuations conducted from ships or boats are included.

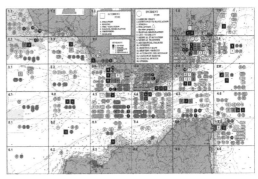

Figure 5. Geographical distribution real emergencies according to subtype. Strait of Gibraltar. Years 2000-2004

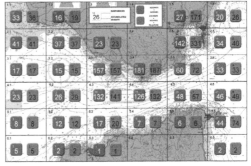

Figure 6. Geographical distribution real emergencies according to severity. Strait of Gibraltar. Years 2000-2004

REFERENCES.

Arroyo, I. Legislación marítima y fuentes complementarias. Tecnos. Madrid (2004). Instrumento de adhesión de 29 de enero de 1993 de España al Convenio de 27 de abril de 1979 Internacional sobre búsqueda y salvamento marítimo, hecho en Hamburgo. BOE nº 103 de 30 de abril de 1993. Madrid (1993).

Instrumento de ratificación de la Convención de las Naciones Unidas sobre el derecho del mar, hecho en Montego Bay el 10 de diciembre de 1982. BOE nº 39 de 14 de febrero de 1997. Madrid (1997).

International Aeronautical and Maritime Search and Rescue Manual. IAMSAR. Volume I. Organization and Management. IMO / ICAO. London / Montreal (1999).

International Aeronautical and Maritime Search and Rescue Manual. IAMSAR. Volume II. Mission Coordination. IMO / ICAO. London / Montreal (2007).

International Convention for Safety of Life at Sea SOLAS. Consolidated Edition. IMO. London (2004).

International Convention for Safety of Life at Sea SOLAS. Amendments of 2003, 2004 and 2005. IMO. London (2006).

23. Safety at Sea – a Review of Norwegian Activities

T. E. Berg & B. Kvamstad
Norwegian Marine Technology Research Institute, Trondheim, Norway

F. Kjersem
Norwegian Coastal Administration

ABSTRACT: This paper offers a brief review of recent and ongoing Norwegian activities aimed at improving safety at sea, focusing primarily on waters under Norwegian administration. Our discussion of international activities is mostly limited to ongoing IMO Polar Code efforts and Arctic Council search and rescue topics.

1 BACKGROUND

Information from insurance companies and class societies shows that there is rising trend in the number of shipping incidents and accidents. Statistics for vessels in Norwegian waters and vessels sailing under the Norwegian flag confirm this tendency. Data from the Norwegian Coastal Administration list annual 80 – 100 groundings in Norwegian waters. Capsizes and sinkings of small fishing vessels are numerous and result in many losses of life at sea in Norwegian waters. In view of the forecasted increase in traffic volume (Arctic Council, 2009), (Ministry of the Environment, 2006) related to Arctic oil and gas production, export of ores and other minerals and possible commercial use of the Northern Sea Route, the Norwegian authorities have identified a need for greater efforts to raise the level of safety for shipping in Norwegian waters, particularly in the north. This paper first discusses current safety levels for shipping in Norwegian waters using the available statistics on maritime incidents and accidents. Some cases will be described to illustrate situations in which contingency planning resources were able to prevent serious accidents. The "Full City" and "Langeland" incidents in July 2009 are taken as examples of emergency response resources being inadequate to prevent major disasters.

Section 3 focuses on the outcomes of a Norwegian pilot study of reducing the risk of shipping accidents in Norwegian waters. This study is being managed by MARINTEK on behalf of the Norwegian Coastal Administration and the Executive Committee for Northern Norway. The outcome of this study will provide guidelines for further Norwegian work on safety at sea.

Section 4 discusses follow-up activities in Northern Norway, bilateral collaboration related to the Barents Sea region and the preparation of a Northern North Atlantic collaborative project on safety at sea, which will involve partners from Norway, Iceland, the Faroe Islands, Greenland, Canada and USA.

The final part of the paper relates Norwegian activities to ongoing IMO Polar Code work and initiatives taken by Arctic Council to improve search and rescue operations in the Arctic.

2 CURRENT SAFETY- LEVELS FOR SHIPPING IN NORWEGIAN WATERS

2.1 Accident statistics

The pilot project "A holistic approach to safety at sea" reviewed sources for information on marine accidents in Norwegian waters (Berg, Kjersem and Kvamstad, 2010). One important source of information is the statistics prepared by the Norwegian Maritime Administration. Figure 1 provides an overview, with the data divided into three classes: groundings, collision with quays and other types of marine casualties. As can be seen there has been a slight increase in both groundings and quay collisions in the course of the past five years. Information on maritime accidents in Norwegian waters information has also been collated by insurance companies (CEFOR), Norwegian Sea Rescue, the Accident Investigation Board Norway (http://www.aibn.no/marine/reports), Telenor Maritime Radio, the Governor of Svalbard and the Petroleum Safety Authority Norway (accidents involving the offshore petroleum sector). When we compare Norwegian and international statistics, it turns out to be difficult to obtain information about the underlying causes of accidents in most reporting schemes. The pilot project concluded that more effort should be put into drawing up better reporting schemes. The reports of the Accident Investigation Board Norway incorporated learning elements that could be of general interest for shipping companies.

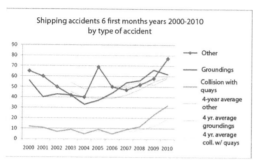

Figure 1. Ship accidents first 6 months 2000 – 2010 by accident types (courtesy of Norwegian Maritime Administration).

2.2 Findings of accidents analysis

Indepth investigations by the Accident Investigation Board Norway provide more information on accident causes and make specific safety recommendations with a view to improving safety at sea based on their findings. These recommendations frequently point to lack of proper crew training, inefficient and/or unclear bridge management, lack of electronic chart system updates and unsatisfactory operating procedures.

2.3 Some illustrative cases

The cargo vessel MV Full City grounded at the Sastein anchorage in Telemark, southern Norway, in the early morning of 31 July 2009 (Figure 2). After arriving at the anchorage anchors were dropped at the position shown in Figure 2. Five shackles (137 m) of the starboard anchor chain were in water when the master reported back to Brevik VTS. The anchor position was 0.9 nautical miles from the nearest shore at a depth of approximately 20 – 22 m. At the time of anchoring the wind was near gale force from the southwest and it deteriorated during the evening. The nearest weather stations showed severe gale to storm with highest measured wind speeds of 48 – 54 knots. Hindcast calculations indicated a wave height of 6 – 8 m south of Sastein. Around midnight the ship's officers observed that the ship was drifting. It has later been estimated that the vessel was drifting at 2 – 3 knots and was less than 0.3 nautical miles from land when the engine was started. Some six minutes after the engine was started the ship

grounded, the engine room was flooded and the main engine stopped. Later investigations showed that both flukes of the anchor had broken off when the vessel was dragging anchor.

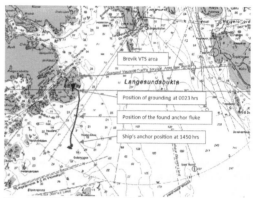

Figure 2. MV Full City grounding during a summer storm in July 2009 (Courtesy of Accident Investigation Board, Norway).

The small cargo vessel MV Langeland got into difficulties in the same storm, en route from Moss in Norway to Karlshavn in Sweden (Figure 3).It was heading for shelter in the Swedish Koster Fiord when communication with the vessel was lost. The Swedish authorities mounted a search and rescue operation. After some time, lifevests and the emergency beacon were found floating in the water. A ROV from the Swedish Navy found the sunken vessel in 108 m of water. Six men lost their lives in this accident.

Figure 3. Planned sailing route for MV Langeland

Another illustrative case is the collision followed by grounding of the fishing vessel Hundvåkøy (March 2010). The vessel is a 1443 GRT, 69 m Norwegian purse seiner, which was sailing southbound through Finnsnesrenna in Northern Norway, fully loaded with 1,400 tonnes of capelin, when it met Hordafor 4 in the narrow passage between the Island of Senja and the mainland (Figure 3). Hordafor 4 is a cargo ship of 65 m, transporting fish feed and raw materials for fish farms along the coast. The vessels collided at 8 am and Hundvåkøy suffered major hull damage that flooded the engine room. This produced a significant starboard list and the vessel was grounded some 500 m south of Finnsnes Bridge. The crew abandoned ship and were picked up by a passing fishing vessel. No injuries resulted from the accident. A prompt inspection by the Norwegian Coastal Administration resulted in a decision to patch the large hole in the hull with a Miko plaster. Both of the Coastal Administration vessels attending the incident, M/V North Crusader and M/V Harstad, were carrying a Miko Salvage Kit from which it was possible to select a six- by two-metre "hat-shaped" patch and, in less than six hours, secure it to the hull to cover the hole. As a result, virtually all of the catch was saved and transferred to other vessels, and the engine room was pumped dry. This enabled the Hundvåkøy to be refloated and towed about 15 nm to the NATO quay at Sørreisa for more permanent repairs.

Salvage experts who attended the incident believe that without the patch, the trawler would almost certainly have capsized and sunk. This would have resulted in a costly recovery operation and a risk of pollution in a sensitive Arctic region

The last example concerns an offshore towing operation. During the return tow of the Hutton platform legs (Figure 5) from Murmansk to Europe the tow entered encountered harsh weather off the coast of Finnmark in Northern Norway. Towing speed was reduced to around 1 knot against a head sea. After some time the towline broke and one of the Norwegian Emergency Towing Vessels operated by the government was called to assist the drifting unit. They were able to connect an emergency towing line and helped the towing vessel to bring the offshore structure into Honningsvåg to reinforce its sea fastenings before the tow was allowed to continue south along the Norwegian coast.

Figure 5. Towing of Hutton leg structure into port of Honningsvåg for reinforcement of sea fastenings (courtesy Ulf Klevstad, Norwegian Coastal Administration).

3 OUTCOMES FROM A PILOT STUDY OF SAFETY AT SEA

In response to questions on safety at sea in Norwegian waters, MARINTEK was requested by the Maritime Forum North organisation to lead a pilot project that would take a holistic view of safety at sea. The project was sponsored by the Executive Committee for Northern Norway and the Norwegian Coastal Administration. A reference group representing governmental bodies, non-governmental organisations and commercial stake-holders was set up to gather information on known issues of safety at sea and new challenges that could be related to a forecast increase in petroleum and shipping activity in the High North. The study listed five themes that needed to be studied:
- Increased understanding of and insight in accidents at sea
- Surveillance and monitoring of traffic at sea
- National and international efforts on safety at sea
- Contingency planning resources and infrastructure

Figure 4. Hundvåkøy accident (courtesy of Norwegian Coastal Administration).

– Training and education, and increased competence in maritime sector and contingency planning organisations

Some of the main challenges identified for marine operations in the High North were the lack of high-quality charts, lower quality of metocean forecasts compared to North Sea standards, reduced communication availability and lack of emergency response infrastructure. Table 1 lists some of the challenges identified under the five topics.

Table 1. Identified challenges - safety at sea

Topic	Challenges
Increased understanding of and insight in accidents at sea	Today's available statistics do not identify the root of causes for the accidents. Incidents and near accidents which could lead to serious situations are not registered in public statistics.
Surveillance and monitoring of traffic at sea	Communication and tracking systems in the Arctic do not offer sufficient performance and capacity to meet todays and future communication and surveillance requirements. Information exchange often redundant and ineffective. In some cases actors involved in e.g. a search and rescue (SAR) operation owns information which could be of use for other actors who are unaware of its existence.
National and international efforts on safety at sea	Changes in international rules and regulations for safety at sea are mainly made by the International Maritime Organisation (IMO). However due to e.g. difference in national interests the decision-making process is slow. There are many recommendations and guidelines for Arctic operations, but very few mandatory regulations.
Contingency planning resources and infra-structure	Excessive response time and too few contingency planning resources present a crucial challenge in the Arctic. Response time is extremely important due to low temperatures and harsh weather conditions. There is potential for raising the level of service by optimizing the available emergency preparedness resources and infrastructure. Important information for vessels at sea is often poor, such as ocean, meteorological, hydrological and bathymetric data. Poor information input is a poor basis for decision-making.
Training and education, and increased competence in maritime sector and contingency planning organisations	There are no institutions that educate SAR personnel with specific competence in operations in harsh arctic weather conditions. Recruiting is difficult, young people chose other professions. Inexperienced personnel may be incapable of coping with a stress situation.

On the basis of the studies performed and the challenges identified, a detailed list of recommendations for future work and research topics was drawn up. The following list summarises the main recommendations and topics which ought to be followed up after the end of the project:

– Establish a national (Norwegian) forum for working with emergency related issues and improved safety at sea. This forum should be closely associated with the Norwegian Coastal Administration's competence centre for contingency planning, safety at sea and surveillance.
– Further development, strengthening and modification of contingency planning infrastructures in order to meet future rises in maritime activities in the Arctic.
– Improve the interaction between private and public contingency planning resources.
– Improve the quality of ocean, meteorological and hydrological data for the fishing and merchant fleets.

Figure 6 shows the one-hour sailing time coverage of the Norwegian coast by rescue vessels operated by Norwegian Sea Rescue: some areas of the northeastern coast lack coverage by these vessels. However, the Coastal Administration has chartered emergency response vessels for Northern Norway.

Figure 6. Coverage of vessels operated by Norwegian Rescue (one-hour sailing time). Red areas: Covered by continuously manned rescue vessel. Yellow areas: Covered by voluntarily manned rescue vessel (courtesy of Norwegian Sea Rescue).

4 FOLLOW-UP WORK ON SAFETY AT SEA IN HIGH NORTH WATERS

4.1 Further work in the MARSAFENORTH project

Maritime Safety Management in the High North (MarSafe North) is an ongoing safety related project where MARINTEK is project manager. Kongsberg Seatex is project owner and the project is financially supported by the Research Council of Norway. It will be finalized in autumn 2011. The project aims to improve safety at sea by identifying users and user needs in transport and advanced marine operations in the shipping and petroleum sectors. The project has also identified technologies for communication, navigation, tracking and surveillance that are capable of meetingcurrent and future user requirements. MarSafe North has identified a number of challenges for marine operations in the High North which will be the basis for recommendations for future efforts. The final phase of the project will include work on ice drift measurement and modeling, communication system analysis, information modeling and feedback to relevant national and international organizations and institutions

4.2 Barents Secretariat project

In collaboration with Maritime Forum North, the Norwegian Coastal Administration and the research company Ocean Futures, MARINTEK has obtained support from the Norwegian Barents Secretariat to start a collaborative pilot project with Russian Federation stakeholders on safety at sea in the Barents Sea. The first workshop was held in Kirkenes in mid-January 2011. The aim of the workshop was to define the state of the art and specify areas in which improvements to safety at sea could be obtained. Potential topics for future collaboration included:
– Preparing data, information, knowledge and expertise for the new Knowledge Centre for Emergency Response and Oilspill Contingency Planning that has been established in Vardø by the Norwegian Coastal Administration
– Improving search and rescue capabilities in the Barents Sea
– Providing input to IMO's efforts on the new Polar Code, especially dedicated competence requirements and the development of training tools for safe operation of ships in the Barents Sea region.
The second project meeting will be held in the Russian Federation in March/April 2011.
Figure 7 shows SAR areas covered by shore-based helicopters in Northern Norway. The Sea King helicopters are operated by the Royal Norwegian Air Force. They have passed their original expected lifetime and the procurement process for new SAR helicopters has been initiated. Norwegian SAR helicopters have been involved in marine rescue operations in the Russian sector of the Barents Sea.

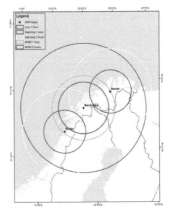

Estimated SAR coverage

Figure 7. Estimated SAR coverage with shore based helicopters in Northern Norway (Courtesy of Rescue Coordination Centre, Bodø)

4.3 Nordic Atlantic Cooperation project

The Nordic Atlantic Cooperation (NORA) has approved a project on safety at sea in the Northern North Atlantic region. MARINTEK is leading a consortium of partners from Norway, the Faroes, Iceland, Greenland, Canada and the USA. A revised project description will be drawn up in early 2011 and the first project meeting is scheduled to take place in May. The main areas of focus for this project will be:
– Improving operational competence for personnel involved in proactive and reactive emergency response activities
– Optimising the use of the existing emergency response infrastructure across national borders
– Reducing response time to reaching a ship in distress in the Northern North Atlantic (as for the North Atlantic Bridge route from Churchill (Canada) to Murmansk (Russian Federation)
Part of the work of this collaborative project will be based on the outcomes of the Arctic Council working group on search and rescue in the Arctic region; see section 5.2 for more information..

4.4 RCN pilot project

In collaboration with Maritime Forum North and the research organisation Ocean Futures, MARINTEK has asked for support for a pilot project to prepare a major research proposal for the MAROFF research programme (Research Council of Norway, RCN) on safety at sea in the High North. The main topics in this proposal will be:
– Collaborative activities on knowledgesharing for High North emergency response operations
– Additional competence requirements for seafarers operating in polar waters

– Specification of transnational infrastructure for search and rescue operations in polar waters.

The main project proposal will be delivered to RCN in mid-February 2011

5 NORWAY'S PART IN INTERNATIONAL SAFETY AT SEA EFFORTS

5.1 IMO's Polar Code

Norway plays an important part in IMO's ongoing efforts to draw up an updated Polar Code. At the 54th session of the Sub-Committee on Ship Design and Equipment, a working group under Norwegian chairmanship worked on the development of a mandatory code for ships operating in polar waters. It has been agreed to utilise a risk-based/goal-based approach including defined goals and functional requirements supported by prescriptive provisions where necessary. A correspondence group was set up to provide input to DE 55, which will take place in March 2011. The main activity of the correspondence group was to review the hazard matrix presented in the DE 54 document (IMO 2010 a). The report of the correspondence group was finalized in mid-December 2010 and forwarded to IMO as a paper for DE 55 (IMO 2010 b). The revised hazards matrix comprises four main elements: Environmental conditions, High latitudes, Environmental sensitivity and Human element. The report recommends that a new working group should be established at DE 55. Guidance has been requested from the DE Sub-Committee on how to understand the concept of "additional hazards" in the context of the Code to be developed. It is assumed that the consequences of any hazard will have to be investigated to determine the potential gaps in the current Conventions and Codes to obtain the same safety level for shipping in polar waters as in other international waters.

5.2 Arctic Council's Search and Rescue efforts – based on publish information and phone calls with Norwegian delegation leader

The Arctic Council Ministerial Meeting in Tromsø (April 2009) decided to establish a task force "to develop and complete negotiations by the next Ministerial Meeting in 2011 of an international instrument on cooperation in search and rescue operations in the Arctic". The USA and the Russian Federation co-chair this task force (Arctic Council, 2010). Work is under way to develop a legally binding document. The task force is at the stage of full intergovernmental negotiations in which "nothing is agreed until everything is agreed. Even though the most complex issues are left to the very end, the co-chairs plan to prepare the agreement for

signature at the Nuuk (Greenland) meeting in May 2011.

6 CONCLUDING REMARKS AND RECOMMENDATION FOR FUTURE WORK

Safety at sea in Arctic waters introduces extra challenges. Remoteness and lack of contingency planning infrastructure are among the causes of some of these additional challenges. It is therefore important that Arctic and Antarctic coastal nations should improve surveillance of all types of marine traffic in their waters in parallel with improving infrastructure and drawing up bilateral agreements for training and infrastructure use. As the consequences of an accident may be more severe in Arctic waters the frequency of incidents must be reduced to keep safety at sea at a given level.

It is important to support the IMO's ongoing efforts to develop a mandatory Polar Code. The next DE55 meeting in March 2011 will discuss the hazard matrix, and it is possible that there will be a new correspondence group after this meeting. Norway is prepared to take a leading role in this work.

The ongoing Arctic Council task force on search and rescue collaboration is scheduled to deliver its draft for a legally binding document at the Ministerial meeting in May 2011, specifying the responsibilities for SAR operations in Arctic waters. However, it is important to note that operators in the Arctic region must take precautions to reduce the level of risk of their operations by reducing the frequency of accidents and mitigating the consequences of accidents that do happen.

On the basis of ongoing work in Norway we recommend further work on:
– Improving regional collaboration to in order optimize the efficiency of existing arctic contingency planning resources
– Developing additional qualification requirements for seafarers operating in arctic waters
– Developing specific training offers combined with knowledge transfer from personnel with cold-climate shipping experience
– Strengthening meeting places for knowledgesharing for senior officers operating ships in arctic waters.

ACKNOWLEDGEMENTS

MARINTEK thanks the Norwegian Coastal Administration, the Executive Committee for Northern Norway and the Barents Secretariat for their financial support for several projects that provided the background for this paper. The assistance of Captain Tor Husjord of Maritime

Forum North is greatly appreciated for his work to establish stakeholder groups for Norwegian projects on safety at sea.

REFERENCES

Arctic Council, 2009: Arctic Marine Shipping Assessment – 2009 Report, PAME, Borgir, Nordurslod - 600 Akureyri, Iceland, 2009

Arctic Council, 2010: Meeting of Senior Arctic Officials, sec. 6.3 Search and Rescue Task Force, AC-SAO-OCT10_FINAL REPORT, Torhavn, Faroe Islands, October 2010

Berg, T. E., Kjersem, F. and Kvamstad, B. 2010. Sjøsikkerhet – en felles utfordring, MARINTEK, Trondheim, Norway, November 2010.

IMO DE54, 2010: Report of the Working Group on Development of a Mandatory Polar Code, DE 54/WP.3, London, UK, October 2010.

IMO DE 55, 2010: Development of a Mandatory Code for Ships Operating in Polar Waters – Report of the Correspondence Group, DE 55/12/**, London, UK, December 2010.

Ministry of the Environment, 2006: A holistic management plan for the marine environment in the Barents Sea and waters outside Lofoten, Governmental White Paper no. 8 (2005-2006), Oslo, Norway.

24. Improving Emergency Supply System to Ensure Port City Safety

Z. Wang, Z. Zhu & W. Cheng

Zhenjiang Watercraft College, Jiangsu, Zhengjiang, China

ABSTRACT: To efficiently implement emergency response program to port unexpected incidents, a perfect emergency supply system which including communications supplying, transportation supplying and rescue equipments supplying must be ready-to-used. Considering physical geography, harbor area, possible incident type and incident scale and other factors, using multi-objective fuzzy decision theory to set up emergency supply centers, improving sharing resource mechanism, administrative legislation and other measures are be used to improve port emergency supply system. Shanghai port's practice prove the improve port emergency supply system is effective.

1 INTRODUCTION

In the last two decades, the boom in international trade and the function expanding of port, have led to more vessels to get in and out port water. There has been a strong focus on the relationship between port water unexpected incidents and port authorities, vessels, berth operators, onboard and port workers, terminal operators, owner and operators of different transport modes interacting with the port water area (rail, road, inland navigation). Port water incident is a kind of serious disaster with high consequence. A disaster at port water is an accident which affects the vessel, the berth, the persons on board or berth, the cargo or the environment.

Marine traffic risk has been a core subject in maritime studies, because it is coupled with transport safety, shipping efficiency, distribution reliability and loss prevention (Tsz Leung Yip, 2006). Some coastal countries have undertaken the task of equipping their coastline with the appropriate sea rescue means, following the guidelines of the International Maritime Organization and in accordance with the International Convention of Search and Rescue (IMO, 1974, 1999). Meanwhile, some decision-making methods have also been used in the sea shipping areas. Port water area is much different from sea water in navigation density, width of fairway, type and number of vessels, and other navigation environment. Meanwhile, vessels' berth and anchorage operations, port operations, port and waterway engineering operations, can lead port water accidents.

The port water incidents to which this work refers are those which occur in port water region. The port water region is the water area within the port bound lines. It includes berths' connecting water areas, port fairways, vessel turn around areas and port anchorages.

Although the port authorities do their best to improve the port water's navigational environment, port city safety is still faced with port operation accidents, maritime accidents, and result in personnel injury or death, property loss, as well as severe environmental damages. If effective rescue operation cannot be taken immediately, a great ecological or economic tragedy is unavoidable. Therefore, the port water incident rescue is an emergency operation that requires quick response.

From the technical viewpoint, port water rescue can be defined as an external action aimed at rescuing persons, vessels, public property, and protecting water area environment from an incident which is immediate and irreversible unless action is taken.

The main activity of the port water emergency rescue services is to attend to incidents that occur in the port water environment. These are services related with the activation of extinguishing a fire, the withdrawal of floating and sinking objects which are dangerous for navigation, separating the unsafe vessel, controlling and clearing up the leakage matters, medical rescue.

When an incident occurs, the authorities in charge the port water area send all of the corresponding reports to the government centre of the port city and to the superior port management department. With this data, official statistics are drawn up, using various parameters related to the characteristics of the vessel and the berth, type of accident and damage incurred. The distribution of the port water rescue resources is planned by using this incident data and the characteristics of the possible locations.

The response time is critical in rescue operations and the rescue charge is very high. The port water rescue activity must be carry out immediately at the time that the emergent rescue centre accepts the official reports. The emergent rescue plan made under time limiting pressure may be not a good one and delay the rescue. The process of planning sea rescue resources and their distribution in the various locations should be carried out according to scientific criteria, both of a technical nature and in terms of cost-effectiveness.

The paper first provides an analysis of the incidents type in the port water area and the relevant rescue resources. Then outline the multi-objective fuzzy decision method based on timeliness and economical efficiency target. At last, describes a practical example of the distribution of a rescue resource and draw some conclusions on the method proposed.

2 INCIDENT TYPE AND THE NECESSARY RESCUE RESOURCES OF PORT WATER AREA

2.1 Critical characteristics of the port water area

Vessels passing in and out the water area, vessels' berthing operations, vessels' anchoring operations, port and waterway engineering operations, make the port water area is the busiest navigation area, and bring the port water area's safety faced with more menaces. Port water area's navigation environment is associated with a number of critical characteristics:

- High traffic volumes;
- Wide variation in vessel size and types;
- High portions of speed craft and ferries;
- Close proximity of marine facilities within a small geographic area;
- A high proportion of coastal and inland water craft;
- Active mid-steam operations for cargo movements;
- Lots of port and waterway engineering operations;
- Multiple water approaches to the port and
- Lots of anchoring areas.

Above characteristics are disadvantageous in port water area. The following are several advantageous key characteristics:

- Good navigation traffic control of vessel activity;
- Perfect navigation system and

- A high level of reporting of port water area incidents.

2.2 Incident types in the port water area

The causes of port water area incident can be various, such as design defects of vessel, navigation traffic events, port operation accident, port and waterway engineering operation failure, badly weather and other disadvantageous factors. The direct impacts of the disaster may include oil/chemical spills, raw sewage discharges, cargo losing, fairway blocks, port paralysis, crews and workers casualties, with indirect impacts such as ecology disaster.

Table 1 illustrates the accident types and the possible consequences of each kind of accident.

Table 1 Incident types and the possible consequences

Type of incidents	Possible consequences
Vessel accidents	Fire/explosion
	Danger cargo falling
	Gas/liquid leakage
	Collision/contract
	Stranding/grounding
	Foundering/sinking
	Capsized/list
	Man overboard
	Structural failure
	Others(e.g., machinery failure)
Port operating accidents	Fire/explosion
	Danger cargo operation failure
	Gas/liquid leakage
	Damage to equipment
	Capsized/list
	Damage to wharf
Heavy nature damage	Fire/explosion
	Danger cargo falling
	Gas/liquid leakage
	Damage to equipment
	Capsized/list
	Damage to vessel
	Damage to wharf

Each kind of incident needs the suitable rescue resources to effectively carry out emergency response.

2.3 The point of determining how to response the incident

Port water incident rescue gains great attention from decision-makers and researchers. On reporting an incident, it always covers the precise location, the latest condition, and the physical parameters (type, construction, and size) of the disaster vessel, the berth, the fairway, the nature and quantity of the cargo, as well as the nature of the damage, in particular any harmful and poisonous substances. Therefore, port water area incident rescue is a necessary and complicated subject.

At the point of determining how to response the port water incident, some factors must be taken into account which is decisive in the process of distributing port water rescue resources. These factors consist of:

1 Characteristics of the incident, the wharf, the fairway, the vessel and the damage produced.
2 Types of the incidents and establishment of a scale of severity.
3 Risk source and sensitive area analysis.
4 Distribution of resources such as tug-boats, fire-fighting boat, oil-absorbing ship, rescue boats, cleaner vessel, helicopters, with a definition of their radius of action.
5 Placement of rescue resources assigning indicators of suitability to every possible incident location.
6 With shorter response time.
7 Cost-effectiveness.

3 METHOD FOR ASSIGNING PORT WATER RESCUE RESOURCES

In the port environment, rescue resources distribution is one of the most important parts of the port emergency response plan. In China, work on the emergency support system is managed by port authorities and the Ministry of Maritime Safety Administration. When the authorities draw up the emergency response plan, they firstly carry out hazards identification and evaluation, and study on sensitive areas according to port function zone.

The efficient assignment of port rescue resource requires, on the one hand, various port operation areas to be interrelated with areas in which accidents are concentrated and on the other hand, a planning process to be developed in which information on the past and present, as well as predictions on the future, are handled. The information is about vessels' information (type, size, performance, loading condition and, manning), port water's navigation environment (traffic flow, hydrological condition, meteorological conditions and, channel condition), and service at port (pilotage service, vessel traffic service, berth condition, and port operations).

Thus, the method to be applied in assigning rescue resources combines aspects of models for the location of port activities with elements of planning, such as port areas planning and port functions planning. A general methodology based on gravitational models to optimize distribute sea rescue resources has been applied to assign 'sea rescue boats' (Azofra, 2007). To overcome limited time pressure, while retaining minimum rescue project duration, a rescue plan for the maritime disaster rescue is obtained with the application of heuristic resource-constrained project scheduling approach (Liang Yan, 2009).

3.1 Problem statement

In the port emergency response plan, according to the characteristic of unexpected incident, the degree and the development, the scale of severity is classified into four grades: particularly serious (Grade I), grave (Grade II), severely (Grade III) and general (Grade IV). The particularly serious incident needs the nation to start the emergency response plan and dispatch domestic each kind of rescue resources, or even receive overseas support resources. The general incident can be deal with the port enterprise by self-prepared resources. The Grade III incident only needs the port authority to use resources of one rescue center. The grave incident needs more resources than one rescue center, so that several rescue centers need to take part in the rescue activity.

Here only the rescue resources distribution of the grave incident is discussed, that is to solve how to use more than two rescue centers to deal with unexpected incident. To the port city, the emergency resources candidate storages are those areas where many berths are concentrated. Port operation areas, port fairway and anchorages are those areas where need resources. To improve port city's safety and competitiveness, the port areas are planed scientifically according to the types of unload cargoes and berthing vessels. Berths with the same function are centralized in one area, so that vessels entering and outgoing the port area are the same. So that the possible incidents in the area are the same and the needed resources are regular. The locations of all of the rescue resources do not necessarily have to coincide and an incident at port water area may or may not require the presence of one or several means of sea rescue.

3.2 Multi-object decision models

The evaluation of location should be performed taking into account several factors of the accidents: their number, type, scale of severity and, considering that any accident should be responded to immediately, the distance between the place where the accidents takes place and the location of the resource to be used. Moreover, for any possible location (port zone, harbor enterprise) its suitability or capacity should be assessed (fire fighting, sewage cleaning and other rescue installations available).

The response time is critical in rescue operations. This will depend on the technical characteristics of the means of rescue used (speed and operability), on the distance to be covered and on the weather conditions at the time of the rescue. However, once the resource to be used is identified, the only controllable parameter is distance. Thus, for the purposes of the present work, it is assumed that time is proportional to distance.

Storage and usage rescue resources have to pay quite a large amount of money. On the one hand, to keep the resource storage into use need a vast cost, this is a flat-rate fee. On the other hand, sending resource from the storage point to demand points need fee which is related to means of delivery and transmission distances. Considering the economic target, the number of storage point to be used should be as small as possible.

The decision of distribution rescue resources is one kind of multi-objective problems. The targets of the decision are that the response time is short and the number of storage point to be used is small.

Provided that one port city has port areas as $B_1, B_2, \cdots, B_p$. The possible needed resources in the port areas are $X_1, X_2, \cdots, X_m$. Supposed that the port has n candidate resource storages $A_1, A_2, \cdots, A_n$. The flat-rate fee of each storage and the weights of the two criterions (timeliness and economic) are known. The weights of the two criterions can be gotten by using Analysis Hierarchy Progress (AHP) according to incident type and characteristic of the port area. The most proper decision is that the rescue resource support scheme based on an overall consideration of timeliness and economic factors. For one kind of incident, the rescue resources' support scheme has two targets as follow:
– The target function of economic

$$C = \sum_{i=1}^{n}\sum_{j=1}^{p} C_{ij} + \sum_{i=1}^{n}\sum_{j=1}^{p} X_{ij} SC_i \qquad (1)$$

where C_{ij} is the costs of resources transmitted from storage $i(i=1,\cdots n)$ to the demand point $j(j=1,\cdots p)$; SC_i is the flat-rate fee of each storage $i(i=1,\cdots n)$. X_{ij} is the criteria be used to judge whether the resource is transmit from storage $i(i=1,\cdots n)$ to the demand point $j(j=1,\cdots p)$. If the resource is transmit for storage i to the demand point j, $X_{ij}=1$, else, $X_{ij}=0$. And there is a relation: $X_{ij}=\text{sgn}(C_{ij})$.
– The target function of timeliness

$$T = \max(\sum_{i=1}^{n}\sum_{j=1}^{p} X_{ij} \times T_{ij}) \qquad (2)$$

where T is the timeliness target. T_{ij} is the time that the resource send form storage $i(i=1,\cdots n)$ to the demand point $j(j=1,\cdots p)$. If there are more than one storage to attend the rescue activity, the timeliness target is the time of the scheme with the longest time.
– The general objective
Use the Linear Weighted Technique (LWT) to get the general objective:

$$S = \min(\lambda_1 T + \lambda_2 C) \qquad (3)$$

where $\lambda_1 + \lambda_2 = 1, \lambda_1 > 0, \lambda_2 > 0$. λ_1, λ_2 can be get by using Uncertainty AHP (WANG Zesheng 2007) based on incident type and incident degree.

Form the function (1) and (2), it is easy to see that the decision variable is X_{ij}.

4 EXAMPLE OF THE MULTI-OBJECT DECISION MODELS: ASSIGNING A RESCUE RESOURCE

The Port of Shanghai faces the East China Sea to the east, and Hangzhou Bay to the south. It includes the heads of the Yangtze River, Huangpu River (which enters the Yangtze River), and Qiantang River. The Port of Shanghai is a critically important transport hub for the Yangtze River region and the most important gateway for foreign trade. The port of Shanghai includes 5 major working zones as shown in Fig 1.

The possible main types of accidents occurring in each zone are shown in Table 2.

Table 2 Port zone area and possible accidents

Port zone area	Possible accidents
Zone A	Collision/ Man overboard
Zone B	Collision/ Grounding
Zone C	Collision/ Damage to wharf
Zone D	Collision/ Damage to equipment
Zone E	Fire/explosion/Gas/liquid leakage

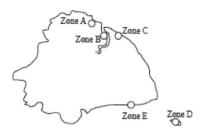

Fig 1 Port zone area

The distances between every zone are shown in Table 3.

Table 3 Distances between every zone (kilometer)

Zone	A	B	C	D	E
A	0	10	30	150	100
B	10	0	15	140	90
C	30	15	0	120	70
D	150	140	120	0	50
E	100	90	70	50	0

The supporting provided that every zone area is the candidate storage, the flat-rate fees of each storage are as shown in Table 4.

Table 4 The flat-rate fee of each storage (ten thousand Yuan RMB)

Zone	A	B	C	D	E
Fee	20	20	20	25	23

The costs of one kind of resource (for example for liquid leakage) transmitted form zone $i(i = 1, \cdots 5)$ to zone $j(j = 1, \cdots 5)$ are shown in Table 4.

Table 5 Transmitting costs between zones (ten thousand Yuan RMB)

Zone	A	B	C	D	E
A	0	0.3	0.5	2	1.5
B	0.3	0	0.4	1.8	1.2
C	0.5	0.4	0	1.7	1.1
D	2	1.8	1.7	0	1
E	1.5	1.2	1.1	1	0

To response liquid leakage accident in port areas, the timeliness is very strong. Set up $\lambda_1 = 0.8, \lambda_2 = 0.2$, by using the above multi-objective model, get that Zone A and Zone E are the best point to set up storage.

5 CONCLUSION

The principle on which a port city should take decisions about the supply center of port rescue resources should be based on highly objective criteria. At present, when assigning rescue resources, a wide variety of technical factors are taken into account (e.g. water area of the port to be covered, traffic flows and types of traffic, danger involved in these traffics and port operations, accident rates, port facilities). By using these factors it is possible to weigh the suitability of a candidate location. In fact, however, political factors often affect and the process of determining the supply center of rescue resources.

Although it is the most subjective factor, it usually has the greatest weight in taking the final decision.

When assigning port rescue resources, it is necessary to use a holistic viewpoint of the problem. This vision is essential if the problem involves management of port water rescue resources. In practice, the management of all port rescue resources of a port city is interdependent, regardless of how they are distributed.

The proposed model is for individual allocation of port rescue resources supply center. The method presupposes that these resources supply center will be assigned one by one, but this does not imply that a previous assignment will not condition a later one. Therefore, once a given resource has been assigned, it must be considered in an interdependent way with the other resources located along the same zone of the port. In this context, it should be pointed out that the assignment of a resource to a location would lead to a reduction of the suitability factor of the rest of the possible locations near the selected location.

REFERENCES

IMO, 1974. International Convention for the Safety of Life at Sea (SOLAS).

IMO, 1999. International Convention on Maritime Search and Rescue.

Tsz Leung Yip, 2006. Port traffic risks-A study of accidents in Hong Kong waters. Transportation Research Part E 44 (2008) 921-931.

Liang Yan, 2009. A heuristic project scheduling approach for quick response to maritime disaster rescue. International Journal of Project Management 27, 620-628.

Azofra M., 2007. Optimum placement of sea rescue resources. Safety Science 45,941-951.

WANG Zesheng, JIN Yongxing, ZHU Zhonghua, 2008. Risk Evaluation of Fairway Traffic Environment Using Uncertainty AHP . The First International Conference on Risk Analysis and Crisis Response.

25. Congested Area Detection and Projection – the User's Requirements

T. Stupak & S. Żurkiewicz
Gdynia Maritime University, Gdynia, Poland

ABSTRACT: Shanghai Maritime University and Gdynia Maritime University have decided to join forces on research program titled "Online detecting and publishing of congested zones at sea". This research seems to be in harmony with e-Navigation concept presented by IALA. The authors have attempted to identify the Congested Area Detection and Projection System user's requirements.

1 BACKGROUND

A number of institutions worldwide are taking actions to assure, improve and promote safety of navigation and marine environment protection. Some of them focus on legal issues, others on technical standards and improvements of equipment quality, communication systems or seafarers training standards.

These actions are usually not well coordinated, whilst connection of individual elements in one coherent system, might result in positive influence on reaching the goals: safety of navigation and marine environment protection.

One of the international institutions which see the need of integration of presently separately operating systems or its elements is International Association of Marine Aids to Navigation and Lighthouse Authorities, (IALA). IALA has made an afford to designing of complete system known as e-Navigation.

In short, one may say, that IALA defines e-Navigation as gathering, integration, exchange and projection of information on board of the ship and shore station alike.

The main tasks of IALA e-Navigation concept are following:
- to improve safety of navigation by providing hydrographic data, navigational warnings and information,
- to improve communication, data and information exchange,
- to secure transport and improve logistics hinterland effectiveness,
- to support effective rescue action in distress, and to collect and archive relevant data for future investigation,

- to integrate and present information on board and on shore,
- to assure global range of coherent norms, compatibility and interoperability of equipment, devices, systems, operational procedures and symbols used,
- to assure adequate accuracy, integrity and continuity so the system could be considered as safe in critical conditions,
- to minimize number of autonomous ship and shore based systems,
- to facilitate effective utilization of waterways for vessels of different classes.

It is highly probable, that elaboration and implementation of uniform standards for e-Navigation will take considerable period of time. Meanwhile, the presently operating systems supporting safe navigation will continue to develop. This considers also the marine traffic monitoring research conducted by Gdynia Maritime University from several years. This research seems to be in harmony with e-Navigation concept presented by IALA.

2 SMU-GMU JOINT PROJECT

2.1 *Project Introduction*

Recently, Shanghai Maritime University and Gdynia Maritime University have decided to join forces on research program titled "Online detecting and publishing of congested zones at sea".

This scientific co-operation is based on Agreement between the Government of the Republic of Poland and the Government of the People's Republic of China about technical and scientific co-operation, signed in Beijing on 13th of April 1995.

2.2 Marine Traffic Monitoring System

In accordance to IMO requirements, maritime states, including China and Poland have introduced Marine Traffic Monitoring Systems utilizing shore based and ships based devices of automatic identification, (AIS). This system facilitates the real time movement monitoring of the vessels fitted in devices of AIS class A or B. AIS class A devices (dedicated for ships), transmits with variable frequency information which include: geographical position, vector of movement as well as declared draught, port of destination and voyage plan as option. Data declared are entered manually by navigating officers. Class B devices (dedicated for fishing vessels and small crafts), transmits with much lesser frequency, information about position and vector. Declared information is not available in this system. In the case of Poland, implementation of such system had been required by Copenhagen Declaration and EC Directive 2002/59, similarly as for remaining Baltic Sea EU member states.

At present marine traffic monitoring is conducted on the Polish waters by the passive mode, and concentrated mainly on data recording. The full use of obtained information by processing them, making available for other users and presenting in useful form, in accordance with e-Navigation concept, may have positive impact on safety of navigation and marine environment protection.

2.3 Marine Safety Information Exchange System

In addition to Marine Traffic Monitoring System, Poland, similarly to the other EU maritime states, has implemented Marine Safety Information Exchanging System in accordance with IMO COM-SAR/Circ.15, form 9th of March 1998, utilizing AIS PL network. This system facilitates transmission by VTS operators of current marine safety information and local warnings to the ships by means of AIS base station. Polish AIS system, (AIS-PL), consists of 11 land based stations (8 marine and 3 inland) linked via a national server to HELCOM network. Although stations spatial distribution was designed to broadcast VHF signals in A1 zone, the whole Polish responsibility area is not permanently covered (Fig. 1).

Required levels of system performance were established in order to satisfy traffic surveillance and maritime safety requests. There is evidence that effective AIS coverage depends on propagation conditions due to weather and pressure. However, anomalous propagation which results in extended VHF range is relatively rare, there are days when single station range increases from 35 to 200 miles and opposite side of Baltic is accessible. Major traffic regions, like VTS Zatoka Gdanska and VTMS Zatoka Pomorska were designed to have extra coverage redundancy in case of system outages or poor propagation. For that purpose there are alternative base stations and additional communication links.

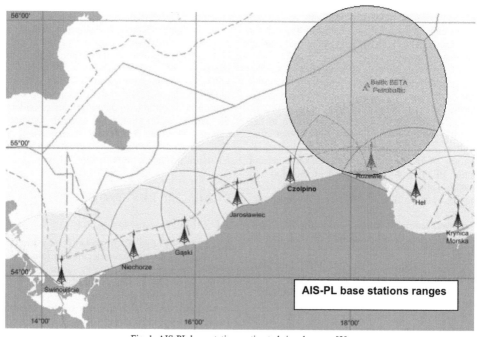

Fig. 1. AIS-PL base stations estimated signal ranges [3]

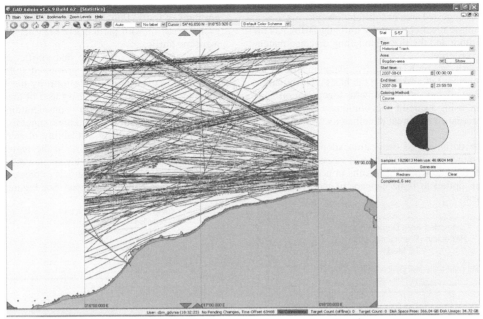

Fig.2. South Baltic Sea traffic flow pattern recorded by AIS PL form 1st to 7th August 2007 [2]

2.4 Information available to Marine Traffic Monitoring System users

The scope of AIS information obtained to ship in relation to those available by VTS stations is basically the same. However, number of ships providing such information is different in both of the cases. This difference in number is caused both by limit of ships transmission range particularly of AIS class B devices, as well as by additional VTS stations possibilities to utilize own system of receiving antennas and to obtain additional AIS information available from the other VTS stations.

VTS centres have access to wider information which is covering bigger area than ships have. National Marine Safety System in Poland is currently under reconstruction and additional information from cameras, and shore radar stations would be available to VTS.

The Figure 2 demonstrates current possibilities of obtaining and recording traffic information by AIS PL system.

Data recorded by AIS PL might be used for purpose of joint SMU-GMU project.

In distinct to VTS stations, the ships might observe in close distance the objects not fitted in AIS devices, which might remain to be not visible for cameras and shore based radar stations. Therefore possibilities of exchanging information are important as safety of navigation and environment protection is concern.

2.5 SMU-GMU joint project description

The main goal of joint SMU-GMU research is to design identification, prediction, and real time projection algorithmic models of congested area.

The figure 3 presents basic idea of the system, as proposed by the authors of this paper.

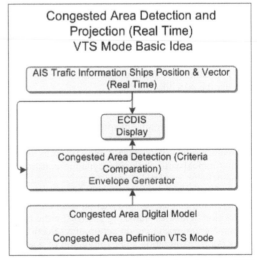

Figure 3. Congested Area Detection and Projection – VTS Mode, Basic Idea

One of the first scientific problems will concern defining and building of mathematical model of congested areas. Due to specific traffic flow pattern on Chinese and Polish waters, each team intends to define congested area in a different way:

– Chinese research team, by utilizing DBSCAN algorithm and the fuzzy modelling distance matching criterion,
– Polish research team, by planning routes of movement of vessels, on the basis of information received from the Automatic Identification System, using a variant of evolutionary algorithms, in accordance with treat of collision criterion.

The definition of congestion adopted by Polish team might require prediction of vessel position in a longer time horizon. This results in need to add prediction option to basic concept of the system. See figure 4.

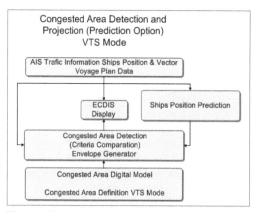

Figure 4. Congested Area Detection and Projection – VTS Mode, Prediction Option

Following tasks will to be performed by GMU research team:

1 Conduct surveys on the movement of vessels in Polish maritime areas and development of:
– Procedures for on-line tracking of vessels under the AIS information.
– Assumptions algorithm modelling areas with particularly high density of maritime traffic through the planning of routes of movement of ships.
2 Create models of passage routes of ships in an environment with static and dynamic constraints and develop and implement alternative evolutionary algorithms for planning paths of movement of vessels on the basis of information from the AIS.
3 Develop heuristics analyzing changes in the environment and reasoning on action, the selection of operators, exploration and exploitation of a set of solutions.
4 Investigate the simulation and test variants of evolutionary algorithms of traffic flow on the basis of different variants of information from the AIS in Polish maritime areas.

5 Analyze and elaborate results of evolutionary algorithms of traffic flow and identification of congested areas or with particularly high density of maritime traffic.

The GMU project provides that traffic information data would be obtained, processed and projected on the screen in a useable form by the VTS station and VTS would be a primary user of the system. However taking in to consideration e-Navigation requirements, the authors suggest, that system should also facilitate availability of selected information addressed to all ships in the region as well as information addressed to the single vessel.

This would require that system has to process data in both, VTS and Own Ship mode. See Figure 5.

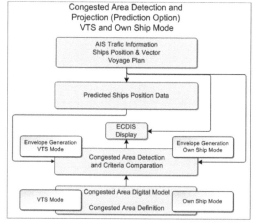

Figure 5. Congested Area Detection and Projection – VTS & Own Ship Mode.

The system for relying of processed data and information would provide for displaying of congested area on the navigation bridge in a real time, as defined according to selected criteria , as well as the early warnings of the collision treats, which might appear in the future on the selected region or single vessel track.

3 USER'S REQUIREMENTS

Regardless of the aims of the SMU and GMU research teams in connection to detection and projection of congested area, the authors have attempt to identify the system user's requirements, both VTS operators and the navigating officers. Taking in to consideration technical feasibility and scientific potential of the partners, these requirements are possible to be met in the frame of the joint project.

3.1 VTS operators requirements

The authors have identified following VTS operators requirements:

Feasibility of detection and projection of the congested area on which, due to the traffic intensity, its flow is reduced or constrained.

Feasibility of detection and projection of congested area on which, traffic intensity exceeds safe margins required to facilitate safe navigation, taking in to account current hydro-metrological conditions.

Feasibility of early detection of collision treat for a single vessel or common treat for group of ships and presentation of potential collision area with indication of time and number of ships involved.

Feasibility of short term prediction of changes in location, shape, range and character of congested area, and in case of collision impended area prediction of time and number of vessels involved.

Detected area in course of time may smoothly change its location, shrink, enlarge, change character, divide, merge or vanish. In the future, the system operators basing on above changes may learn to draw conclusions concerning congestion level and navigation safety.

Early detection of area impendent by the congestion feasibility might be restricted in the case of lack of voyage plan transmission by class A ships. This data transmission is not obligatory at the moment. Therefore authors postulate obligatory voyage plan transmission on the area covered by VTS, on the same base as obligatory VHF reports are required at present.

The voyage plan information covering controlled area obtained from all ships equipped in AIS class A, would facilitate preparation of reliable forecasts of areas impendent by congestion and would facilitate effective traffic management.

Scientific analyze of recorded changes of congested area might open new prospects in long term forecasting of congestion or collision treats. This analyze would contribute to design new methods of congested area traffic management and elaboration of the new, presently unknown navigation aids, dedicated to voyage planning in the form of charts, modelled on Pilot or Routing Charts.

3.2 Navigating officers' requirements

The authors have identified following navigating officer's requirements:

1 Feasibility of Internet access to real time presentation of ship's movements in selected VTS area including detected congestion boarders and short term prediction of the following:
   - area of slow or constrained traffic flow,
   - area with traffic intensity exciding safe margins,
   - area impended by collision treat.
2 Feasibility of receiving facsimile transmission of information described in point 1, in equal time intervals, for example every one hour.
3 Feasibility of obtaining automatically, or by an action of VTS operator of early SMS warning about impendent situation on the ships track, boarders of suspected area and time of reaching these boarders.
4 Feasibility of receiving automatically SMS message about treat diminishing due to own vessel early action or action of the other vessels involved.

The authors anticipate that both teams would analyze identified in this paper user's requirements and consider to implement them in the course of research.

If presented proposals would meet the interest of research teams, the authors are ready to engage in preparation of detailed specification of the user's requirements which might facilitate including such requirements in the joint project.

4 CONCLUSION

Implementation on the laboratory level of SMS early warnings or messages related to collision impendent area, after conducting required tests might in the future be developed in the direction of Automatic Collision Impendent Area Avoidance Advisory System.

As the next step, the scope of the automatic advices might be extended to Congested Area Avoidance System.

Automatic massages related to congestion avoidance might compose an element of that future Congested Area Traffic Management System, which seems to be in harmony with IALA e-Navigation concept.

Authors express view, that present joint SMU-GMU project could be a great opportunity as a start point for future fruitful cooperation that would benefit of all seafarers, marine environment and both nations.

REFERENCES

[1] Król A., Stupak T: Dokładność rejestracji danych pozycyjnych statków w systemie nawigacji zintegrowanej. XIV International Conference Transcomp 2010, Logistyka 6/2010 str. 1675 – 1681
[2] Duda D., Stupak T., Wawruch R.: "Ship movement and tracking with AIS". Polish Journal of environmental studies, Vol. 16, No 3C, 2007, pp. 18-25.
[3] www.umgdy.pl

26. Studying Probability of Ship Arrival of Yangshan Port with AIS (Automatic Identification System)

H. Y. Ni Ni
Myanmar Maritime University, Yangon, Myanmar
Shanghai Maritime University, Shanghai, China
Q. Hu & C. Shi
Shanghai Maritime University, Shanghai, China

ABSTRACT: The distribution pattern is considered to be a poission distribution for periodical schedule.The evolution of the ship arrival distribution patterns and the χ^2 fit test for observation are based on the ship dynamic data of international harbour in Yangshan. AIS (Automatic Identification System) is used the frequency of ship arrivals in this study .This study aims to implement the test for performance of ship arrival distributions and theirs probabilities. The ship arrival distribution in the spread sheet simulation systems was found to follow poission distribution; its frequency distribution is changed by observation system, tends to change the system's of the probabilities.

1 INTRODUCTION

The ship arrival distribution test is a vital piece basic research for port planning and for the choice of distribution pattern in the simulation approach. The test implements the ship arrival distribution approach. The test result of the ship arrival distribution will influence the choice of port queuing system ,which subsequently influences relevant variables measured from such model .The purpose of this paper is to find the ship arrival distribution and their probabilites in the spread sheet simulation systems. After many ship arrivals are merged and verification are required to define the statistical pattern of the ship arrival time distribution. Base on ship dynamic data of international port in Yangshan and using AIS data for this study. This study also investigated the evolution of the ship arrival distribution pattern and theirs probabilities by observation system .It was focused on the probability distribution pattern of the arrival distribution for vessels with the poission distribution.

2 OPERATION BACKGROUNG

Vessel Traffic studies are necessary in harbour construction. In planning and design of harbours and use both real vessel and scale model to do experiment and collect data. Also depends on statistical studies and synthesize domestic marine traffic data. Ships routeing system is at the sea area Yangshan port. The Yangshan deep-water port is a new port in Hangzhou Bay south of Shanghai. 27.5 kilometres from Shanghai's southern coast, and under the juris diction of the neighbouring Province Zhejiang, was chosen as the site of the deepwater port of Shanghai. The average water depth in the area of the islands is over 15 meters. Yangshan deep water port has five container berths, each around 15 meters deep.

AIS is excepted to play a major role in ship reporting system .The systems is typically included in the static voyage related and dynamic data automatically provided by the AIS system .The use of the AIS long range feature, where information is exchanged via communications satellite ,may be implemented to satisfy the requirements of ship reporting systems . AIS will play a role in overall international maritime information system, supporting voyage planning and monitoring. This will assist administrations to monitor all the vessels in their areas of concern and tracks.

Figure 1 Yangshan Deepwater Port in China

3 MODEL DEVELOPMENT

The scope of present study considers the effect of harbour allocation on arriving times.Therefore, necessary to model the system starting from the ship arrivals and theirs probabilities to the berth operations.This simulation model was developed spread sheet Excel software.

3.1 *Ship's arrival model*

The ships arrive at a datum line randomly, the number of ships arriving at the datum line in a given interval of times is a random variable and its distribution fits the poission distribution (k.Hara, 1966), and the probability is:

$$P(X = k) = \frac{\lambda^k e^{-\lambda}}{k!} \quad k = 0, 1, 2, 3, \ldots \quad (1)$$

where $P(X=k)$ = Probability that k ship will arrived at the daum line in a given interval of time t; λ = average number of ships arriving at the datum line in unit time ; e = base of Naperian logarithm, e= 2.718; t = given interval of time.

If no ship will arrive in the time interval t, that is k=0, then

$$P(X = 0) = \frac{\lambda^0 e^{-\lambda}}{0!} = e^{-\lambda} \quad (2)$$

The distribution of the number of ships arriving enterend into a harbour in a week Figure.2 and The data of the daily number of ships entered into a harbour in a week are fits the Poission Distribution in Figure.3.

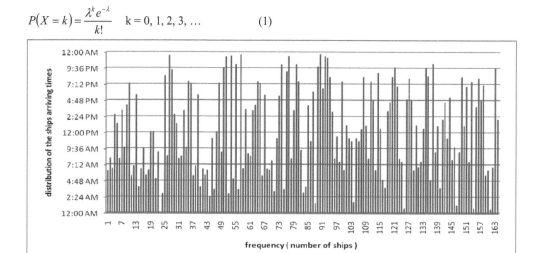

Figure 2 The distribution of the number of ships arriving

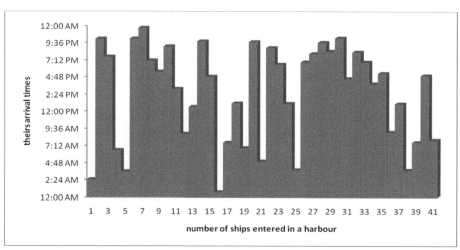

Figure 3 Number of ships entered in a harbour and theris arrival times

196

3.2 *Discussion of test approch*

First, a base model is developed and AIS data of ship arrivals are used the frequency of ships arrivals. The data consists distribution of 164 ships arrivals a period of a week .The generated data is used to run actually arrived at the port in a week.The empirical frequency distribution of daily number of ship is sorted out and fitted the Poission Distribution in Table 1.

$$\lambda = \frac{\text{the total number of ships in a week}}{\text{the number of days in a week}} = \frac{41}{7} = 5.84 ships/day$$

Table 1 The Empirical frequency distribution of Daily Number of ships

n	f_j	Frequency (f_j/N)
1	1	0.143
2	1	0.143
3	0	-
4	1	0.143
5	0	-
6	1	0.143
7	0	-
8	1	0.143
9	1	0.143
10	0	-
11	1	0.143
Total	7	1.001

3.3 χ^2 *fit test*

The most appropriate approaches are the χ^2 fit test. χ^2 fit test is to be applied .Hypothesis, the empirical frequency distribution of the daily number of ships fits the Poission Distribution.

$$\chi^2 = \sum_{j=1}^{g} \frac{f_j - F_j}{F_j} \qquad (3)$$

where f_j= the frequency of group j for empirical distribution; F_j = the frequency of group j for Poission Distribution, $F_j = NP_j$; N = the volume of a week; P_j = the probability of Poisson Distribution; g= the number of groups

Test the sample in this study should be divided into group before applying this approach to the ship arrival distribution. Grouping (selecting the number of groups and group arrival) is critical factor for the χ^2 fit test in this study. The χ^2 fit test of statistical analysis is applied for the ship arrival distribution test in this study.

3.4 *Simulation*

After simulating the base model, a replication was performed for each of the ships arriving times ranging for a week before any data is recorded to sure state has been achieved .The model is then run for another year to obtain the annual throughput.

4 RESULT

The purpose of the present study is to understand the various ships arrival on the times within a week and their probabilities. Table.2 shows the process of calculation for χ^2 and figure.4 shows the arriving groups and their probabilities percentage.

The result of the calculation is χ^2 = 10.637. DF =g-γ-1= 7-1-1=5 (γ is the number of parameter of the poission Distribution), α =0.05. The Table of χ^2 Distribution (Table.3) is referenced.

χ^2 α = 11.070. Owing to the fact that $\chi^2 < \chi^2$ α , the hypothesis cannnot be rejected. That is to say, there is no real evidence to doubt that the empirical frequency distribution of the daily number of ships fits the Poission Distribution.

Table 2 χ^2 Caluculation

n	f_j	P_j	F_j	χ^2
1	1	0.016	0.112	7.040
2	1	0.049	0.343	1.258
3	0	-	-	-
4	1	0.140	0.98	0.001
5	0	-	-	-
6	1	1.160	1.12	0.012
7 and more	3	0.035	0.245	2.326
Total	7			10.637

Table 3 χ^2 Distribution Table

DF			α		
	0.10	0.05	0.025	0.01	0.001
1	2.706	3.814	5.024	6.635	10.828
2	4.605	5.991	7.378	9.210	13.816
3	6.251	7.815	9.348	11.345	16.266
4	7.779	9.488	11.143	13.277	18.467
5	9.236	11.070	12.833	15.086	20.515
6	10.645	12.592	14.449	16.812	22.458
7	12.017	14.067	16.013	18.475	24.322

5 CONCLUSION

The ship arrival distribution varies depending on the test approaches of the number of group interval. The χ^2 fit test is hard to pass with larger samples, the study suggested that the threadhold limit value should be modified appropriately for larger samples in order to conform to realistic needs .The result provides the statistical analysis of ships' arrival times and their probabilities at Yangshan terminal in China .

Figure 4 The arriving groups and their probabilities percentage

REFERENCES

Zun Jun. 2006. Marine Traffic Engineering. Dalian Maritime University Published.

Tu-Cheng Kuo,Wen_Chih Huang, Sheng_Chieh Wu and pei_Lun Cheng. 2006. A case Study of inter-arrival Time Distributions Of Container Ship. Journal of marine science and Techonology (14)3: 155-164.

Chi Zhi-gang, Liu Xiao-dong and Zhao Yue-qiong. 2008. Research on Probability Distribution of Container Liner's Arrival. Proceeding of the IEEE.

Shen Ai-di, Han-fang, Victoria.2008. AIS ship-based station in Zhoushan Marine information monitoring system in Hong Kong. (29)3: 10-13.

Kyay Mone S.O., HU Q., SHI C., WEINTRIT A, Gdynia Maritime University, Gdynia 2010. *Clustering Analysis and Identification of Marine Traffic Congested Zones at Wusongkou, Shanghai. Zeszyty Naukowe Akademii Morskiej w Gdyni, Scientific Booklet No 67, Problemy nawigacji i transportu morskiego.*

Yang Mu & Lionel Ho. Date of Publication: 22 June 2006. Shangahi's Yangshan Deep Water Port: An International Mega Port in the making. EAI Background Brief No.290.

On Corretions to the Chi-squared Distribution.Journal of the Royal Statistical Society Series B (Methodological).1958 (20)2: 387.

JohnE.kuhn,MS,MD,MaryLouV.H.,Greenfield,MPH,MS,and Edward M.Wojtys,MD.1997 . A Statistics primer Statistical Tests for Discrete Data.The American Journal of sports medicine, (25)4.

T.Tengku_Adnan,D.Sier,R.N.Ibrahim.2009. Performance of ship Queuing Rules at Coal Export Terminals . Proceeding of the IEEE.

J.MEDHI, Stochastic Model in Queuing Theory. second edition.

Author index